Introduction to Agronomy

Introduction to Agronomy

Edited by
Cassius Foster

Larsen & Keller
www.larsen-keller.com

Introduction to Agronomy
Edited by Cassius Foster
ISBN: 978-1-63549-020-6 (Hardback)

⊟ Larsen & Keller

Published by Larsen and Keller Education,
5 Penn Plaza,
19th Floor,
New York, NY 10001, USA

Cataloging-in-Publication Data

Introduction to agronomy / edited by Cassius Foster.
    p. cm.
Includes bibliographical references and index.
ISBN 978-1-63549-020-6
1. Agronomy. 2. Agriculture. I. Foster, Cassius.
SB91 .I68 2017
631--dc23

The publisher's policy is to use permanent paper from mills that operate a sustainable forestry policy. Furthermore, the publisher ensures that the text paper and cover boards used have met acceptable environmental accreditation standards.

Printed and bound in the United States of America.

For more information regarding Larsen and Keller Education and its products, please visit the publisher's website www.larsen-keller.com

Table of Contents

Permissions

Index

Preface

Agronomy is that branch of agriculture, which deals with the study of plants in order to manufacture food, fiber, fuel, etc. The major topics included in agronomy are plant breeding, crop rotation, soil classification, weed control, irrigation, plant physiology, soil fertility, pest control, etc. Contents in this book focus on the practices that promote the production of healthier food. This book provides detailed information about the subject of agronomy. It aims to understand the various topics that fall under this discipline, and also examines the techniques and how these have practical applications. Coherent flow of topics and student-friendly language make this text an invaluable source of knowledge.

A detailed account of the significant topics covered in this book is provided below:

Chapter 1- Agronomy is the science concerned with holistic production and management of agricultural products. It seeks to provide high-quality food stuff while minimising damage done to the environment. Agronomy is an emerging field of study; the following chapter will not only provide an overview, it will also delve deep into the variegated topics related to it.

Chapter 2- Agronomy is an aggregation of knowledges that are derived from other fields. Some of these are agrophysics, agroecology, nutrient management and food systems. Agronomy constantly studies the inputs required for crop production. This chapter visits these as well as other themes. The major components of agronomy are discussed in this chapter.

Chapter 3- Agronomy focuses on efficient techniques of cultivation, pest control and plant breeding. It also aims to aims to protect the soil through the cultivation of cover crops. These practices change according to region as the themes like shifting cultivation and crop rotation will show. Organic farming is a major branch of agronomic practice.

Chapter 4- Tillage is one of the oldest methods of agriculture that is still in practice. Tillage involves the preparation of the soil before planting. Techniques of tillage alter according to the nature of the soil, its depth and drainage. Some techniques such as no-till farming, strip till farming and mulch-till farming are also explained in this chapter.

Chapter 5- Commercialisation of agriculture has led to great progress in the fields of plant modification, high-yielding seeds and agricultural tools. Agriculture is an important part of the economy and is still witnessing vast amounts of progress and development. Some themes in this chapter include the green revolution, extensive farming and intensive farming.

Chapter 6- Weeding and pest management are a very important part of agronomic practice. Agricultural pests do great harm to agricultural produce. The chapter touches upon pesticides, herbicides and weed control. The chapter discusses the methods of agronomy in a critical manner providing key analysis to the subject matter.

Chapter 7- An alternative form of farming, permaculture aims to provide solutions to the problem of overexploitation of the land as well as the use of insecticides. Permaculture seeks to work alongside nature to design solutions for better cultivation and to also re-invest into land any surplus that is gained from agricultural practice. This chapter will provide an integrated understanding of permaculture.

Chapter 8- Harvesting is a labour-intensive process and requires skill, acumen and effort. It is only natural then that the harvest is one of the stages that would see high amounts of mechanisation and automation of its various processes. Some of the themes discussed in this chapter are on reapers, threshing methods, combine harvesters and the practice of winnowing.

Chapter 9- Agriculture on the whole depends on natural and man-made systems of irrigation. The creation of canals and aqueducts has been going on for centuries. Irrigation is very important for farming. Some methods discussed in this chapter include surface irrigation and drip irrigation. The major categories of irrigation are dealt with in great details in the chapter.

It gives me an immense pleasure to thank our entire team for their efforts. Finally in the end, I would like to thank my family and colleagues who have been a great source of inspiration and support.

Editor

An Introduction to Agronomy

Agronomy is the science concerned with holistic production and management of agricultural products. It seeks to provide high-quality food stuff while minimising damage done to the environment. Agronomy is an emerging field of study; the following chapter will not only provide an overview, it will also delve deep into the variegated topics related to it.

Agronomy (*agrós* 'field' + *nómos* 'law') is the science and technology of producing and using plants for food, fuel, fiber, and land reclamation. Agronomy has come to encompass work in the areas of plant genetics, plant physiology, meteorology, and soil science. It is the application of a combination of sciences like biology, chemistry, economics, ecology, earth science, and genetics. Agronomists of today are involved with many issues, including producing food, creating healthier food, managing the environmental impact of agriculture, and extracting energy from plants. Agronomists often specialise in areas such as crop rotation, irrigation and drainage, plant breeding, plant physiology, soil classification, soil fertility, weed control, and insect and pest control.

Plant Breeding

An agronomist field sampling a trial plot of flax.

This area of agronomy involves selective breeding of plants to produce the best crops under various conditions. Plant breeding has increased crop yields and has improved the nutritional value of numerous crops, including corn, soybeans, and wheat. It has also led to the development of new types of plants. For example, a hybrid grain called triticale was produced by crossbreeding rye and wheat. Triticale contains more usable protein than does either rye or wheat. Agronomy has also been instrumental in fruit and vegetable production research.

Biotechnology

Purdue University agronomy professor George Van Scoyoc explains the difference between forest and prairie soils to soldiers of the Indiana National Guard's Agribusiness Development Team at the Beck Agricultural Center in West Lafayette, Indiana

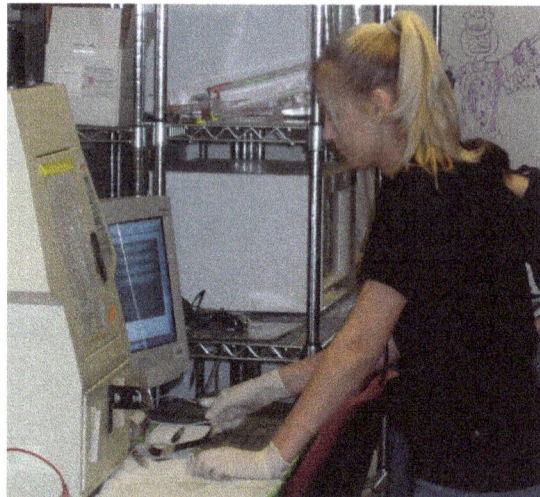

An agronomist mapping a plant genome

Agronomists use biotechnology to extend and expedite the development of desired characteristic. Biotechnology is often a lab activity requiring field testing of the new crop varieties that are developed.

In addition to increasing crop yields agronomic biotechnology is increasingly being applied for novel uses other than food. For example, oilseed is at present used mainly for margarine and other food oils, but it can be modified to produce fatty acids for detergents, substitute fuels and petrochemicals.

Soil Science

Agronomists study sustainable ways to make soils more productive and profitable. They classify soils and analyze them to determine whether they contain nutrients vital to plant growth. Common macronutrients analyzed include compounds of nitrogen, phosphorus, potassium, calcium, magnesium, and sulfur. Soil is also assessed for several micronutrients, like zinc and boron. The percentage of organic matter, soil pH, and nutrient holding capacity (cation exchange capacity) are tested in a regional laboratory. Agronomists will interpret these lab reports and make recommendations to balance soil nutrients for optimal plant growth.

Soil Conservation

In addition, agronomists develop methods to preserve the soil and to decrease the effects of erosion by wind and water. For example, a technique called contour plowing may be used to prevent soil erosion and conserve rainfall. Researchers in agronomy also seek ways to use the soil more effectively in solving other problems. Such problems include the disposal of human and animal manure, water pollution, and pesticide build-up in the soil. Techniques include no-tilling crops, planting of soil-binding grasses along contours on steep slopes, and contour drains of depths up to 1 metre.

Agroecology

Agroecology is the management of agricultural systems with an emphasis on ecological and environmental perspectives. This area is closely associated with work in the areas of sustainable agriculture, organic farming, alternative food systems and the development of alternative cropping systems.

Theoretical Modeling

Agronomy Schools

Agronomy programs, offered at colleges, universities, and specialized agricultural schools, often involve classes across a range of departments, including agriculture, biology, chemistry, economics, and physiology. Completing the coursework usually takes from four to twelve years. Many companies will pay agronomists-in-training's college expenses if they agree to work for them when they graduate.

Integrated Aspects of Agronomy

Agronomy is an aggregation of knowledges that are derived from other fields. Some of these are agrophysics, agroecology, nutrient management and food systems. Agronomy constantly studies the inputs required for crop production. This chapter visits these as well as other themes. The major components of agronomy are discussed in this chapter.

Agricultural Soil Science

Agricultural soil science is a branch of soil science that deals with the study of edaphic conditions as they relate to the production of food and fiber. In this context, it is also a constituent of the field of agronomy and is thus also described as soil agronomy.

History

Prior to the development of pedology in the 19th century, agricultural soil science (or edaphology) was the only branch of soil science. The bias of early soil science toward viewing soils only in terms of their agricultural potential continues to define the soil science profession in both academic and popular settings as of 2006. (Baveye, 2006)

Current Status

Agricultural soil science follows the holistic method. Soil is investigated in relation to and as integral part of terrestrial ecosystems but is also recognized as a manageable natural resource.

Agricultural soil science studies the chemical, physical, biological, and mineralogical composition of soils as they relate to agriculture. Agricultural soil scientists develop methods that will improve the use of soil and increase the production of food and fiber crops. Emphasis continues to grow on the importance of soil sustainability. Soil degradation such as erosion, compaction, lowered fertility, and contamination continue to be serious concerns. They conduct research in irrigation and drainage, tillage, soil classification, plant nutrition, soil fertility, and other areas.

Although maximizing plant (and thus animal) production is a valid goal, sometimes it may come at high cost which can be readily evident (e.g. massive crop disease stemming from monoculture) or long-term (e.g. impact of chemical fertilizers and pesticides on human health). An agricultural soil scientist may come up with a plan that can maximize production using sustainable methods and solutions, and in order to do that he must look into a number of science fields including agricultural science, physics, chemistry, biology, meteorology and geology.

Soil Variables

Some soil variables of special interest to agricultural soil science are:

- Soil texture or soil composition: Soils are composed of solid particles of various sizes. In decreasing order, these particles are sand, silt and clay. Every soil can be classified according to the relative percentage of sand, silt and clay it contains.

- Aeration and porosity: Atmospheric air contains elements such as oxygen, nitrogen, carbon and others. These elements are prerequisites for life on Earth. Particularly, all cells (including root cells) require oxygen to function and if conditions become anaerobic they fail to respire and metabolize. Aeration in this context refers to the mechanisms by which air is delivered to the soil. In natural ecosystems soil aeration is chiefly accomplished through the vibrant activity of the biota. Humans commonly aerate the soil by tilling and plowing, yet such practice may cause degradation. Porosity refers to the air-holding capacity of the soil.

- Drainage: In soils of bad drainage the water delivered through rain or irrigation may pool and stagnate. As a result, prevail anaerobic conditions and plant roots suffocate. Stagnant water also favors plant-attacking water molds. In soils of excess drainage, on the other hand, plants don't get to absorb adequate water and nutrients are washed from the porous medium to end up in groundwater reserves.

- Water content: Without soil moisture there is no transpiration, no growth and plants wilt. Technically, plant cells loose their pressure. Plants contribute directly to soil moisture. For instance, they create a leafy cover that minimizes the evaporative effects of solar radiation. But even when plants or parts of plants die, the decaying plant matter produces a thick organic cover that protects the soil from evaporation, erosion and compaction.

- Water potential: Water potential describes the tendency of the water to flow from one area of the soil to another. While water delivered to the soil surface normally flows downward due to gravity, at some point it meets increased pressure which causes a reverse upward flow. This effect is known as water suction.

- Horizonation: Typically found in advanced and mature soils, horizonation refers to the creation of soil layers with differing characteristics. It affects almost all soil variables.

- Fertility: A fertile soil is one rich in nutrients and organic matter. Modern agricultural methods have rendered much of the arable land infertile. In such cases, soil can no longer support on its own plants with high nutritional demand and thus needs an external source of nutrients. However, there are cases where human activity is thought to be responsible for transforming rather normal soils into super-fertile ones.

- Biota and soil biota: Organisms interact with the soil and contribute to its quality in innumerable ways. Sometimes the nature of interaction may be unclear, yet a rule is becoming evident: The amount and diversity of the biota is "proportional" to the quality of the soil. Clades of interest include bacteria, fungi, nematodes, annelids and arthropods.

- Soil acidity or soil pH and cation-exchange capacity: Root cells act as hydrogen pumps and the surrounding concentration of hydrogen ions affects their ability to absorb nutrients. pH is a measure of this concentration. Each plant species achieves maximum growth in a particular pH range, yet the vast majority of edible plants can grow in soil pH between 5.0 and 7.5.

Soil Fertility

Agricultural soil scientists study ways to make soils more productive. They classify soils and test them to determine whether they contain nutrients vital to plant growth. Such nutritional substances include compounds of nitrogen, phosphorus, and potassium. If a certain soil is deficient in these substances, fertilizers may provide them. Agricultural soil scientists investigate the movement of nutrients through the soil, and the amount of nutrients absorbed by a plant's roots. Agricultural soil scientists also examine the development of roots and their relation to the soil. Some agricultural soil scientists try to understand the structure and function of soils in relation to soil fertility. They grasp the structure of soil as porous solid. The solid frames of soil consist of mineral derived from the rocks and organic matter originated from the dead bodies of various organisms. The pore space of the soil is essential for the soil to become productive. Small pores serve as water reservoir supplying water to plants and other organisms in the soil during the rain-less period. The water in the small pores of soils is not pure water; they call it soil solution. In soil solution, various plant nutrients derived from minerals and organic matters in the soil are there. This is measured through the cation exchange capacity. Large pores serve as water drainage pipe to allow the excessive water pass through the soil, during the heavy rains. They also serve as air tank to supply oxygen to plant roots and other living beings in the soil. In short, agricultural soil scientists see the soil as a vessel, the most precious one for us, containing all of the substances needed by the plants and other living beings on earth.

Soil Preservation

In addition, agricultural soil scientists develop methods to preserve the agricultural productivity of soil and to decrease the effects on productivity of erosion by wind and water. For example, a technique called contour plowing may be used to prevent soil erosion and conserve rainfall. Researchers in agricultural soil science also seek ways to use the soil more effectively in addressing associated challenges. Such challenges include the beneficial reuse of human and animal wastes using agricultural crops; agricultural soil management aspects of preventing water pollution and the build-up in agricultural soil of chemical pesticides.

Employment of Agricultural Soil Scientists

Most agricultural soil scientists are consultants, researchers, or teachers. Many work in the developed world as farm advisors, agricultural experiment stations, federal, state or local government agencies, industrial firms, or universities. Within the USA they may be trained through the USDA's Cooperative Extension Service offices, although other countries may use universities, research institutes or research agencies. Elsewhere, agricultural soil scientists may serve in international organizations such as the Agency for International Development and the Food and Agriculture Organization of the United Nations.

Quotations

[The key objective of the soil science discipline is that of] finding ways to meet growing human needs for food and fiber while maintaining environmental stability and conserving resources for future generations

—*John W. Doran, 2002 SSSA President, 2002*

Many people have the vague notion that soil science is merely a phase of agronomy and deals only with practical soil management for field crops. Whether we like it or not this is the image many have of us

—*Charles E. Kellog, 1961*

Agrophysics

Agrophysics is a branch of science bordering on agronomy and physics, whose objects of study are the agroecosystem - the biological objects, biotope and biocoenosis affected by human activity, studied and described using the methods of physical sciences. Using the achievements of the exact sciences to solve major problems in agriculture, agrophysics involves the study of materials and processes occurring in the production and processing of agricultural crops, with particular emphasis on the condition of the environment and the quality of farming materials and food production.

Agrophysics is closely related to biophysics, but is restricted to the biology of the plants, animals, soil and an atmosphere involved in agricultural activities and biodiversity. It is different from biophysics in having the necessity of taking into account the specific features of biotope and biocoenosis, which involves the knowledge of nutritional science and agroecology, agricultural technology, biotechnology, genetics etc.

The needs of agriculture, concerning the past experience study of the local complex soil and next plant-atmosphere systems, lay at the root of the emergence of a new branch – agrophysics – dealing this with experimental physics. The scope of the branch starting from soil science (physics) and originally limited to the study of relations within the soil environment, expanded over time onto influencing the properties of agricultural crops and produce as foods and raw postharvest materials, and onto the issues of quality, safety and labeling concerns, considered distinct from the field of nutrition for application in food science.

Research centres focused on the development of the agrophysical sciences include the Institute of Agrophysics, Polish Academy of Sciences in Lublin, and the Agrophysical Research Institute, Russian Academy of Sciences in St. Petersburg.

Research institutes and societies

- Agrophysical Research Institute in St. Petersburg, Russia

- Bohdan Dobrzański Institute of Agrophysics in Lublin, Poland

- The Indian Society of AgroPhysics

Scholarly journals

- *Acta Agrophysica*

- *Journal of Agricultural Physics*

- *Polish Journal of Soil Science*

Agroecology

A community-supported agriculture share of crops.

Agroecology is the study of ecological processes that operate in agricultural production systems. The prefix *agro-* refers to *agriculture*. Bringing ecological principles to bear in agroecosystems can suggest novel management approaches that would not otherwise be considered. The term is often used imprecisely and may refer to "a science, a movement, [or] a practice." Agroecologists study a variety of agroecosystems, and the field of agroecology is not associated with any one particular method of farming, whether it be organic, integrated, or conventional; intensive or extensive. Although it has much more common thinking and principles with some of the before mentioned farming systems.

Ecological Strategy

Agroecologists do not unanimously oppose technology or inputs in agriculture but instead assess how, when, and if technology can be used in conjunction with natural, social and human assets. Agroecology proposes a context- or site-specific manner of studying agroecosystems, and as such, it recognizes that there is no universal formula or recipe for the success and maximum well-being of an agroecosystem. Thus, agroecology is not defined by certain management practices, such as the use of natural enemies in place of insecticides, or polyculture in place of monoculture.

Instead, agroecologists may study questions related to the four system properties of agroecosystems: productivity, stability, sustainability and equitability. As opposed to disciplines that are concerned with only one or some of these properties, agroecologists see all four properties as interconnected and integral to the success of an agroecosystem. Recognizing that these properties are found on varying spatial scales, agroecologists do not limit themselves to the study of agroecosystems at any one scale: gene-organism-population-community-ecosystem-landscape-biome, field-farm-community-region-state-country-continent-global.

Agroecologists study these four properties through an interdisciplinary lens, using natural sciences to understand elements of agroecosystems such as soil properties and plant-insect interactions, as well as using social sciences to understand the effects of farming practices on rural communities, economic constraints to developing new production methods, or cultural factors determining farming practices.

Approaches

Agroecologists do not always agree about what agroecology is or should be in the long-term. Different definitions of the term agroecology can be distinguished largely by the specificity with which one defines the term "ecology," as well as the term's potential political connotations. Definitions of agroecology, therefore, may be first grouped according to the specific contexts within which they situate agriculture. Agroecology is defined by the OECD as "the study of the relation of agricultural crops and environment." This definition refers to the "-ecology" part of "agroecology" narrowly as the natural environment. Following this definition, an agroecologist would study agriculture's various relationships with soil health, water quality, air quality, meso- and micro-fauna, surrounding flora, environmental toxins, and other environmental contexts.

A more common definition of the word can be taken from Dalgaard et al., who refer to agroecology as the study of the interactions between plants, animals, humans and the environment within agricultural systems. Consequently, agroecology is inherently multidisciplinary, including factors from agronomy, ecology, sociology, economics and related disciplines. In this case, the "-ecology" portion of "agroecology is defined broadly to include social, cultural, and economic contexts as well. Francis et al. also expand the definition in the same way, but put more emphasis on the notion of food systems.

Agroecology is also defined differently according to geographic location. In the global south, the term often carries overtly political connotations. Such political definitions of the term usually ascribe to it the goals of social and economic justice; special attention, in this case, is often paid to the traditional farming knowledge of indigenous populations. North American and European uses of the term sometimes avoid the inclusion of such overtly political goals. In these cases, agroecology is seen more strictly as a scientific discipline with less specific social goals.

Agro-population Ecology

This approach is derived from the science of ecology primarily based on population ecology, which over the past three decades has been displacing the ecosystems biology of Odum. Buttel explains the main difference between the two categories, saying that "the application of population ecology to agroecology involves the primacy not only of analyzing agroecosystems from the perspective of the population dynamics of their constituent species, and their relationships to climate and biogeochemistry, but also there is a major emphasis placed on the role of genetics."

Inclusive Agroecology

Rather than viewing agroecology as a subset of agriculture, Wojtkowski takes a more encompassing perspective. In this, natural ecology and agroecology are the major headings under ecology. Natural ecology is the study of organisms as they interact with and within natural environments. Correspondingly, agroecology is the basis for the land-use sciences. Here humans are the primary governing force for organisms within planned and managed, mostly terrestrial, environments.

As key headings, natural ecology and agroecology provide the theoretical base for their respective sciences. These theoretical bases overlap but differ in a major way. Economics has no role in the functioning of natural ecosystems whereas economics sets direction and purpose in agroecology.

Under agroecology are the three land-use sciences, agriculture, forestry, and agroforestry. Although these use their plant components in different ways, they share the same theoretical core.

Beyond this, the land-use sciences further subdivide. The subheadings include agronomy, organic farming, traditional agriculture, permaculture, and silviculture. Within this system of subdivisions, agroecology is philosophically neutral. The importance lies in providing a theoretical base hitherto lacking in the land-use sciences. This allows progress in biocomplex agroecosystems including the multi-species plantations of forestry and agroforestry.

Applications

To arrive at a point of view about a particular way of farming, an agroecologist would first seek to understand the contexts in which the farm(s) is(are) involved. Each farm may be inserted in a unique combination of factors or contexts. Each farmer may have their own premises about the meanings of an agricultural endeavor, and these meanings might be different from those of agroecologists. Generally, farmers seek a configuration that is viable in multiple contexts, such as family, financial, technical, political, logistical, market, environmental, spiritual. Agroecologists want to understand the behavior of those who seek livelihoods from plant and animal increase, acknowledging the organization and planning that is required to run a farm.

Views on Organic and non-organic Milk Production

Because organic agriculture proclaims to sustain the health of soils, ecosystems, and people, it has much in common with Agroecology; this does not mean that Agroecology is synonymous with organic agriculture, nor that Agroecology views organic farming as the 'right' way of farming. Also, it is important to point out that there are large differences in organic standards among countries and certifying agencies.

Three of the main areas that agroecologists would look at in farms, would be: the environmental impacts, animal welfare issues, and the social aspects.

Environmental impacts caused by organic and non-organic milk production can vary significantly. For both cases, there are positive and negative environmental consequences.

Compared to conventional milk production, organic milk production tends to have lower eutrophication potential per ton of milk or per hectare of farmland, because it potentially reduces leaching of nitrates (NO_3^-) and phosphates (PO_4^-) due to lower fertilizer application rates. Because organic milk production reduces pesticides utilization, it increases land use per ton of milk due to decreased crop yields per hectare. Mainly due to the lower level of concentrates given to cows in organic herds, organic dairy farms generally produce less milk per cow than conventional dairy farms. Because of the increased use of roughage and the, on-average, lower milk production level per cow, some research has connected organic milk production with increases in the emission of methane.

Animal welfare issues vary among dairy farms and are not necessarily related to the way of producing milk (organically or conventionally).

A key component of animal welfare is freedom to perform their innate (natural) behavior, and this is stated in one of the basic principles of organic agriculture. Also, there are other aspects of animal welfare to be considered - such as freedom from hunger, thirst, discomfort, injury, fear, distress, disease and pain. Because organic standards require loose housing systems, adequate bedding, restrictions on the area of slatted floors, a minimum forage proportion in the ruminant diets, and tend to limit stocking densities both on pasture and in housing for dairy cows, they potentially promote good foot and hoof health. Some studies show lower incidence of placenta retention, milk fever, abomasums displacement and other diseases in organic than in conventional dairy herds. However, the level of infections by parasites in organically managed herds is generally higher than in conventional herds.

Social aspects of dairy enterprises include life quality of farmers, of farm labor, of rural and urban communities, and also includes public health.

Both organic and non-organic farms can have good and bad implications for the life quality of all the different people involved in that food chain. Issues like labor conditions, labor hours and labor rights, for instance, do not depend on the organic/non-organic characteristic of the farm; they can be more related to the socio-economical and cultural situations in which the farm is inserted, instead.

As for the public health or food safety concern, organic foods are intended to be healthy, free of contaminations and free from agents that could cause human diseases. Organic milk is meant to have no chemical residues to consumers, and the restrictions on the use of antibiotics and chemicals in organic food production has the purpose to accomplish this goal. Although dairy cows in both organic and conventional farming practices can be exposed to pathogens, it has been shown that, because antibiotics are not permitted as a preventative measure in organic practices, there are far fewer antibiotic resistant pathogens on organic farms. This dramatically increases the efficacy of antibiotics when/if they are necessary.

In an organic dairy farm, an agroecologist could evaluate the following:

1. Can the farm minimize environmental impacts and increase its level of sustainability, for instance by efficiently increasing the productivity of the animals to minimize waste of feed and of land use?

2. Are there ways to improve the health status of the herd (in the case of organics, by using biological controls, for instance)?

3. Does this way of farming sustain good quality of life for the farmers, their families, rural labor and communities involved?

Views on no-till Farming

No-tillage is one of the components of conservation agriculture practices and is considered more environmental friendly than complete tillage. There is a general consensus that no-till can increase soils capacity of acting as a carbon sink, especially when combined with cover crops.

No-till can contribute to higher soil organic matter and organic carbon content in soils, though reports of no-effects of no-tillage in organic matter and organic carbon soil contents also exist, depending on environmental and crop conditions. In addition, no-till can indirectly reduce CO_2 emissions by decreasing the use of fossil fuels.

Most crops can benefit from the practice of no-till, but not all crops are suitable for complete no-till agriculture. Crops that do not perform well when competing with other plants that grow in untilled soil in their early stages can be best grown by using other conservation tillage practices, like a combination of strip-till with no-till areas. Also, crops which harvestable portion grows underground can have better results with strip-tillage, mainly in soils which are hard for plant roots to penetrate into deeper layers to access water and nutrients.

The benefits provided by no-tillage to predators may lead to larger predator populations, which is a good way to control pests (biological control), but also can facilitate predation of the crop itself. In corn crops, for instance, predation by caterpillars can be higher in no-till than in conventional tillage fields.

In places with rigorous winter, untilled soil can take longer to warm and dry in spring, which may delay planting to less ideal dates. Another factor to be considered is that organic residue from the prior year's crops lying on the surface of untilled fields can provide a favorable environment to pathogens, helping to increase the risk of transmitting diseases to the future crop. And because no-till farming provides good environment for pathogens, insects and weeds, it can lead farmers to a more intensive use of chemicals for pest control. Other disadvantages of no-till include underground rot, low soil temperatures and high moisture.

Based on the balance of these factors, and because each farm has different problems, agroecologists will not atest that only no-till or complete tillage is the right way of farming. Yet, these are not the only possible choices regarding soil preparation, since there are intermediate practices such as strip-till, mulch-till and ridge-till, all of them - just as no-till - categorized as conservation tillage. Agroecologists, then, will evaluate the need of different practices for the contexts in which each farm is inserted.

In a no-till system, an agroecologist could ask the following:

1. Can the farm minimize environmental impacts and increase its level of sustainability; for instance by efficiently increasing the productivity of the crops to minimize land use?

2. Does this way of farming sustain good quality of life for the farmers, their families, rural labor and rural communities involved?

History

Pre-WWII

The notions and ideas relating to crop ecology have been around since at least 1911 when F.H. King released *Farmers of Forty Centuries*. King was one of the pioneers as a proponent of more quantitative methods for characterization of water relations and physical properties of soils. In the late 1920s the attempt to merge agronomy and ecology was born with the development of the field of crop ecology. Crop ecology's main concern was where crops would be best grown. Actually, it was only in 1928 that agronomy and ecology were formally linked by Klages.

The first mention of the term agroecology was in 1928, with the publication of the term by Bensin in 1928. The book of Tischler (1965), was probably the first to be actually titled 'agroecology'. He analysed the different components (plants, animals, soils and climate) and their interactions within an agroecosystem as well as the impact of human agricultural management on these components. Other books dealing with agroecology, but without using the term explicitly were published by the German zoologist Friederichs (1930) with his book on agricultural zoology and related ecological/environmental factors for plant protection, and by American crop physiologist Hansen in 1939 when both used the word as a synonym for the application of ecology within agriculture.

Post-WWII

Gliessman mentions that post-WWII, groups of scientists with ecologists gave more focus to experiments in the natural environment, while agronomists dedicated their attention to the cultivated systems in agriculture. According to Gliessman, the two groups kept their research and interest apart until books and articles using the concept of agroecosystems and the word agroecology started to appear in 1970. Dalgaard explains the different points of view in ecology schools, and the fundamental differences, which set the basis for the development of agroecology. The early ecology school of Henry Gleason investigated plant populations focusing in the hierarchical levels of the organism under study.

Friederich Clement's ecology school, however included the organism in question as well as the higher hierarchical levels in its investigations, a "landscape perspective". However, the ecological schools where the roots of agroecology lie are even broader in nature. The ecology school of Tansley, whose view included both the biotic organism and their environment, is the one from which the concept of agroecosystems emerged in 1974 with Harper.

In the 1960s and 1970s the increasing awareness of how humans manage the landscape and its consequences set the stage for the necessary cross between agronomy and ecology. Even though, in many ways the environmental movement in the US was a product of the times, the Green Decade, spread an environmental awareness of the unintended consequences of changing ecological processes. Works such as *Silent Spring*, and *The Limits to Growth*, and changes in legislation such as the Clean Air Act, Clean Water Act, and the National Environmental Policy Act caused the public to be aware of societal growth patterns, agricultural production, and the overall capacity of the system.

Fusion with Ecology

After the 1970s, when agronomists saw the value of ecology and ecologists began to use the agricultural systems as study plots, studies in agroecology grew more rapidly. Gliessman describes that the innovative work of Prof. Efraim Hernandez X., who developed research based on indigenous systems of knowledge in Mexico, led to education programs in agroecology. In 1977 Prof. Efraim Hernandez X. explained that modern agricultural systems had lost their ecological foundation when socio-economic factors became the only driving force in the food system. The acknowledgement that the socio-economic interactions are indeed one of the fundamental components of any agroecosystems came to light in 1982, with the article Agroecologia del Tropico Americano by Montaldo. The author argues that the socio-economic context cannot be separated from the agricultural systems when designing agricultural practices.

In 1995 Edens et al. in Sustainable Agriculture and Integrated Farming Systems solidified this idea proving his point by devoting special sections to economics of the systems, ecological impacts, and ethics and values in agriculture. Actually, 1985 ended up being a fertile and creative year for the new discipline. For instance in the same year, Miguel Altieri integrated how consolidation of the farms, and cropping systems impact pest populations. In addition, Gliessman highlighted that socio-economic, technological, and ecological components give rise to producer choices of food production systems. These pioneering agroecologists have helped to frame the foundation of what we today consider the interdisciplinary field of agroecology and have led to advances in a number of farming systems. In Asian rice, for example, crop diversification by growing flowering crops in strips beside rice fields has recently been demonstrated to reduce pests so effectively (by the flower nectar attracting and supporting parasitoids and predators) that insecticide spraying is reduced by 70%, yields increase by 5%, together resulting in an economic advantage of 7.5%(Gurr et al., 2016).

Publications

Year	Author(s)	Title
1928	Klages	Crop ecology and ecological crop geography in the agronomic curriculum
1939	Hanson	Ecology in agriculture
1956	Azzi	Agricultural ecology
1965	Tischler	Agrarökologie
1973	Janzen	Tropical agroecosystems
1974	Harper	The need for a focus on agro-ecosystems
1976	Loucks	Emergence of research on agroecosystems
1977	Hernandez Xolocotzi	Agroecosistemas de Mexico
1978	Gliessman	Agroecosistemas y tecnologia agricola tradicional
1979	Hart	Agroecosistemas: conceptos básicos
1979	Cox & Atkins	Agricultural ecology: an analysis of world food production systems
1980	Hart	Agroecosistemas
1981	Gliessman, Garcia & Amador	The ecological basis for the application of traditional agricultural technology in the management of tropical agroecosystems

1982	Montaldo	Agroecologia del trópico americano
1983	Altieri	Agroecology
1984	Lowrance, Stinner & House	Agricultural ecosystems: unifying concepts
1985	Conway	Agroecosystems analysis
1987	Altieri	Agroecology: the scientific basis of alternative agriculture
1990	Allen, Dusen, Lundy, & Gliessman	Integrating social, environmental, and economic issues in sustainable agriculture
1990	Gliessman	Agroecology: researching the ecological basis for sustainable agriculture
1990	Carroll, Vandermeer & Rosset	Agroecology
1990	Altieri & Hecht	Agroecology and small farm development
1991	Caporali	Ecologia per l'agricultura
1991	Bawden	Systems thinking in agriculture
1993	Coscia	Agricultura sostenible
1998	Gliessman	Agroecology: ecological processes in sustainable agriculture
2001	Flora	Interactions between agroecosystems and rural communities
2001	Gliessman	Agroecosystem sustainability
2002	Dalgaard, Porter & Hutchings	Agroecology, scaling, and interdisciplinarity
2003	Francis et al.	Agroecology: The Ecology of Food Systems
2004	Clements, Shrestra	New Dimension in Agroecology
2007	Bland and Bell	A Holon Approach to Agroecology
2007	Gliessman	Agroecology: The Ecology of Sustainable Food Systems
2007	Warner	Agroecology in Action
2009	Wezel, Soldat	A quantitative and qualitative historical analysis of the scientific discipline agroecology
2009	Wezel et al.	Agroecology as a science, a movement or a practice. A review

By Region

The principles of agroecology are expressed differently depending on local ecological and social contexts.

Latin America

Latin America's experiences with North American Green Revolution agricultural techniques have opened space for agroecologists. Traditional or indigenous knowledge represents a wealth of pos-sibility for agroecologists, including "exchange of wisdoms."

Madagascar

Most of the historical farming in Madagascar has been conducted by indigenous peoples. The French colonial period disturbed a very small percentage of land area, and even included some useful experiments in Sustainable forestry. Slash-and-burn techniques, a component of some shifting cultivation systems have been practised by natives in Madagascar for centuries. As of 2006 some of the major agricultural products from slash-and-burn methods are wood, charcoal and grass for Zebu grazing. These practices have taken perhaps the greatest toll on land fertility since the end of French rule, mainly due to overpopulation pressures.

Theoretical Production Ecology

Theoretical production ecology tries to quantitatively study the growth of crops. The plant is treated as a kind of biological factory, which processes light, carbon dioxide, water, and nutrients into harvestable parts. Main parameters kept into consideration are temperature, sunlight, standing crop biomass, plant production distribution, nutrient and water supply.

Modelling

Modelling is essential in theoretical production ecology. Unit of modelling usually is the crop, the assembly of plants per standard surface unit. Analysis results for an individual plant are generalised to the standard surface, e.g. the Leaf Area Index is the projected surface area of all crop leaves above a unit area of ground.

Processes

The usual system of describing plant production divides the plant production process into at least five separate processes, which are influenced by several external parameters.

Two cycles of biochemical reactions constitute the basis of plant production, the light reaction and the dark reaction.

- In the light reaction, sunlight photons are absorbed by chloroplasts which split water into an electron, proton and oxygen radical which is recombined with another radical and released as molecular oxygen. The recombination of the electron with the proton yields the energy carriers NADH and ATP. The rate of this reaction often depends on sunlight intensity, leaf area index, leaf angle and amount of chloroplasts per leaf surface unit. The maximum theoretical gross production rate under optimum growth conditions is approximately 250 kg per hectare per day.

- The dark reaction or Calvin cycle ties atmospheric carbon dioxide and uses NADH and ATP to convert it into sucrose. The available NADH and ATP, as well as temperature and carbon dioxide levels determine the rate of this reaction. Together those two reactions are termed photosynthesis. The rate of photosynthesis is determined by the interaction of a number of factors including temperature, light intensity and carbon dioxide.

- The produced carbohydrates are transported to other plant parts, such as storage organs and converted into secondary products, such as amino acids, lipids, cellulose and other chemicals needed by the plant or used for respiration. Lipids, sugars, cellulose and starch can be produced without extra elements. The conversion of carbohydrates into amino acids and nucleic acids requires nitrogen, phosphorus and sulfur. Chlorophyll production requires magnesium, while several enzymes and coenzymes require trace elements. This means, nutrient supply influences this part of the production chain. Water supply is essential for transport, hence limits this too.

- The production centers, i.e. the leaves, are sources, the storage organs, growth tips or other destinations for the photosynthetic production are sinks. The lack of sinks can be a limiting factor for production too, as happens e.g. in apple orchards where insects or night frost have destroyed the blossoms and the produced assimilates cannot be converted into apples. Biennial and perennial plants employ the stored starch and fats in their storage organs to produce new leaves and shoots the next year.

- The amount of crop biomass and the relative distribution of biomass over leaves, stems, roots and storage organs determines the respiration rate. The amount of biomass in leaves determines the leaf area index, which is important in calculating the gross photosynthetic production.

- extensions to this basic model can include insect and pest damage, intercropping, climatical changes, etc.

Parameters

Important parameters in theoretical production models thus are:

Climate

- Temperature - The temperature determines the speed of respiration and the dark reaction. A high temperature combined with a low intensity of sunlight means a high loss by respiration. A low temperature combined with a high intensity of sunlight means that NADH and ATP heap up but cannot be converted into glucose because the dark reaction cannot process them swiftly enough.

- Light - Light, also called photosynthetic Active Radiation (PAR) is the energy source for green plant growth. PAR powers the light reaction, which provides ATP and NADPH for the conversion of carbon dioxide and water into carbohydrates and molecular oxygen. When temperature, moisture, carbon dioxide and nutrient levels are optimal, light intensity determines maximum production level.

- Carbon dioxide levels - Atmospheric carbon dioxide is the sole carbon source for plants. About half of all proteins in green leaves have the sole purpose of capturing carbon dioxide.

  Although CO_2 levels are constant under natural circumstances [on the contrary, CO2 concentration in the atmosphere has been increasing steadily for 200 years], CO_2 fertilization is common in greenhouses and is known to increase yields by on average 24% [a specific value, e.g., 24%, is meaningless without specification of the "low" and "high" CO2 levels being compared].

C_4 plants like maize and sorghum can achieve a higher yield at high solar radiation intensities, because they prevent the leaking of captured carbon dioxide due of the spatial separation of carbon dioxide capture and carbon dioxide use in the dark reaction. This means that their photorespiration is almost zero. This advantage is sometimes offset by a higher rate of maintenance respiration. In most models for natural crops, carbon dioxide levels are assumed to be constant.

Crop

- Standing crop biomass - Unlimited growth is an exponential process, which means that the amount of biomass determines the production. Because an increased biomass implies higher respiration per surface unit and a limited increase in intercepted light, crop growth is a sigmoid function of crop biomass.

- Plant production distribution - Usually only a fraction of the total plant biomass consists of useful products, e.g. the seeds in pulses and cereals, the tubers in potato and cassava, the leaves in sisal and spinach etc. The yield of usable plant portions will increase when the plant allocates more nutrients to this parts, e.g. the high-yielding varieties of wheat and rice allocate 40% of their biomass into wheat and rice grains, while the traditional varieties achieve only 20%, thus doubling the effective yield.

  Different plant organs have a different respiration rate, e.g. a young leaf has a much higher respiration rate than roots, storage tissues or stems do. There is a distinction between "growth respiration" and "maintenance respiration".

  Sinks, such as developing fruits, need to be present. They are usually represented by a discrete switch, which is turned on after a certain condition, e.g. critical daylength has been met.

Care

- Water supply - Because plants use passive transport to transfer water and nutrients from their roots to the leaves, water supply is essential to growth, even so that water efficiency rates are known for different crops, e.g. 5000 for sugar cane, meaning that each kilogram of produced sugar requires up to 5000 liters of water.

- Nutrient supply - Nutrient supply has a twofold effect on plant growth. A limitation in nutrient supply will limit biomass production as per Liebig's Law of the Minimum. With some crops, several nutrients influence the distribution of plant products in the plants. A nitrogen gift is known to stimulate leaf growth and therefore can work adversely on the yield of crops which are accumulating photosynthesis products in storage organs, such as ripening cereals or fruit-bearing fruit trees.

Phases in Crop Growth

Theoretical production ecology assumes that the growth of common agricultural crops, such as cereals and tubers, usually consists of four (or five) phases:

- Germination - Agronomical research has indicated a temperature dependence of germination time (GT, in days). Each crop has a unique critical temperature (CT, dimension temperature) and temperature sum (dimensions temperature times time), which are related as follows.

$$GT = \frac{TS}{\sum_{k=1}^{N}(T - T_{\text{crit}})}$$

When a crop has a temperature sum of e.g. 150 °C·d and a critical temperature of 10 °C, it will germinate in 15 days when temperature is 20 °C, but in 10 days when temperature is 25 °C. When the temperature sum exceeds the threshold value, the germination process is complete.

- Initial spread - In this phase, the crop does not cover the field yet. The growth of the crop is linearly dependent on leaf area index, which in its turn is linearly dependent on crop biomass. As a result, crop growth in this phase is exponential.

- Total coverage of field - in this phase, growth is assumed to be linearly dependent on incident light and respiration rate, as nearly 100% of all incident light is intercepted. Typically, the Leaf Area Index (LAI) is above two to three in this phase. This phase of vegetative growth ends when the plant gets a certain environmental or internal signal and starts generative growth (as in cereals and pulses) or the storage phase (as in tubers).

- Allocation to storage organs - in this phase, up to 100% of all production is directed to the storage organs. Generally, the leaves are still intact and as a result, gross primary production stays the same. Prolonging this phase, e.g. by careful fertilization, water and pest management results directly in a higher harvest.

- Ripening - in this phase, leaves and other production structures slowly die off. Their carbohydrates and proteins are transported to the storage organs. As a result, the LAI and, hence, the primary production decreases.

Existing Plant Production Models

Plant production models exist in varying levels of scope (cell, physiological, individual plant, crop, geographical region, global) and of generality: the model can be crop-specific or be more generally applicable. In this section the emphasis will be on crop-level based models as the crop is the main area of interest from an agronomical point of view.

As of 2005, several crop production models are in use. The crop growth model SUCROS has been developed during more than 20 years and is based on earlier models. Its latest revision known dates from 1997. The IRRI and Wageningen University more recently developed the rice growth model ORYZA2000. This model is used for modeling rice growth. Both crop growth models are open source. Other more crop-specific plant growth models exist as well.

SUCROS

SUCROS is programmed in the Fortran computer programming language. The model can and has been applied to a variety of weather regimes and crops. Because the source code of Sucros is open source, the model is open to modifications of users with FORTRAN programming experience. The official maintained version of SUCROS comes into two flavours: SUCROS I, which has non-inhibited unlimited crop growth (which means that only solar radiation and temperature determine growth) and SUCROS II, in which crop growth is limited only by water shortage.

ORYZA2000

The ORYZA2000 rice growth model has been developed at the IRRI in cooperation with Wageningen University. This model, too, is programmed in FORTRAN. The scope of this model is limited to rice, which is the main food crop for Asia.

Other Models

The United States Department of Agriculture has sponsored a number of applicable crop growth models for various major US crops, such as cotton, soy bean, wheat and rice. Other widely used models are the precursor of SUCROS (SWATR), CERES, several incarnations of PLANTGRO, SUBSTOR, the FAO-sponsored CROPWAT, AGWATER and the erosion-specific model EPIC., cropping system CropSyst

A less mechanistic growth and competition model, called the Conductance Model, has been developed, mainly at Warwick-HRI, Wellesbourne, UK. This model simulates light interception and growth of individual plants based on the lateral expansion of their crown zone areas. Competition between plants is simulated by a set algorithms related to competition for space and resultant light intercept as the canopy closes. Some versions of the model assume overtopping of some species by others. Although the model cannot take account of water or mineral nutrients, it can simulate individual plant growth, variability in growth within plant communities and inter-species competition.

External Resources

- Wageningen University, Netherlands, Department of Theoretical production Ecology
- Leuven University, Belgium, Department of Theoretical Production Ecology
- Summary page with US government sponsored crop growth models

Agrology

Agrology (*agros*, "field, tilled land"; and *-logia*) is the branch of soil science dealing with the production of crops. The use of the term is most active in Canada. Use of the term outside Canada is sporadic but significant. The term appears especially well established in Russia and China, with agrologists on university faculty lists and agrology curricula.

Agrology is synonymous with agricultural science when used in Canada, is nearly synonymous with the U.S. term *agronomy*, and has a meaning related to agricultural soil science when used outside of Canada.

Canada

The term agrologist was coined by Dr. J. B. Harrington and adopted in 1946 to fill the need in Canada to have a term to denote "provincial agriculturalist". The title of Professional Agrologist is conferred on persons with at least a Bachelor's Degree in Agriculture and who can demonstrate the qualities needed to responsibly teach, practise, or conduct experiments and research in the agricultural sciences. According to the Agricultural Institute of Canada website, an agrologist can also hold a degree in a field related to agriculture, or in some provinces pass rigorous prescribed examinations to attain a professional designation. There are about 5000 agrologists in Canada as of 2004.

Agrology designations are managed by separate governing bodies in each province, with each operating under its own legislation. For example, within British Columbia the term "agrology" is defined by an Act of the Legislature passed in 2003 and adopted in 2004 entitled the Agrologists Act. This Act authorizes the self-governing body, the British Columbia Institute of Agrologists and those practising agrology within British Columbia do so under the following definition.

"Agrology" means using agricultural and natural sciences and agricultural and resource economics, including collecting or analyzing data or carrying out research or assessments, to design, evaluate, advise on, direct or otherwise provide professional support to

(a) the cultivation, production, improvement, processing or marketing of aquatic or terrestrial plants or animals, or

(b) the classification, management, use, conservation, protection, restoration, reclamation or enhancement of aquatic or terrestrial ecosystems that are affected by, sustain, or have the potential to sustain the cultivation or production of aquatic or terrestrial plants or animals;

BCIA, the governing body for Agrologists in British Columbia, has over 1000 registered members.

The Registrars of Professional Agrologists across Canada adopted the following definition of Agrology in May 2007.

Agrology is the practice of bioresource sciences to provide knowledge and advice to support the development of the agriculture sector and the health of the society, environment, and economy.

Outside Canada

Outside of Canada, the term agrology is synonymous with soil science and is not in common usage in English-speaking countries.

Agrology in Soil Science Society Glossaries

Two national member societies (Canadian, American) of the International Union of Soil Sciences (IUSS) maintain and publish glossaries of scientific terms. Other soil science societies defer to the American glossary. The term agrology is not in use. Edaphology or crop edaphology in combination with soil management would be the preferred approach used by soil scientists to concisely describe soil science as it applies to crop production.

Agrology Dictionary Definitions

As of 2004, no dictionary definition of agrology is yet consistent with the Canadian use of the term and dictionary definitions fall into one of four categories.

1. agrology is defined as synonymous with soil science. The root agr- is represented as meaning soil.

2. agrology is defined as synonymous with soil science, but the context implies that soil science is a subdiscipline of agricultural science.

3. agrology is defined as the subdiscipline of soil science as it applies to crop production. This would make agrology synonymous with the term crop edaphology.

4. agrology is defined as the subdiscipline of agronomy that considers the influence of soil.

Phenology

Phenology is the study of periodic plant and animal life cycle events and how these are influenced by seasonal and interannual variations in climate, as well as habitat factors (such as elevation). Examples include the date of emergence of leaves and flowers, the first flight of butterflies and the first appearance of migratory birds, the date of leaf colouring and fall in deciduous trees, the dates of egg-laying of birds and amphibia, or the timing of the developmental cycles of temperate-zone honey bee colonies. In the scientific literature on ecology, the term is used more generally to indicate the time frame for any seasonal biological phenomena, including the dates of last appearance (e.g., the seasonal phenology of a species may be from April through September).

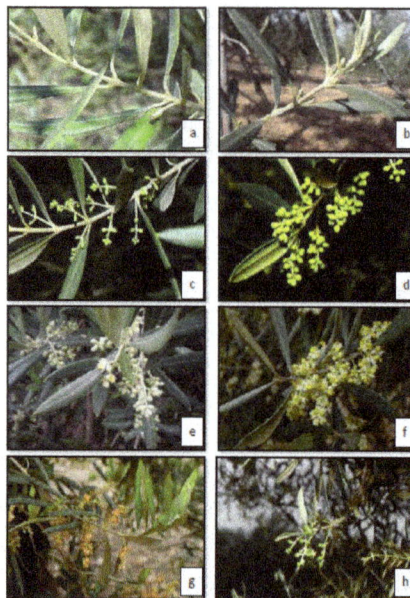

Phenological development of Olive flowering, following BBCH standard scale. a-50, b-51, c-54, d-57, (<15% open flowers); f-65, (>15% open flowers); g-67, (<15% open flowers); h-68 (Oteros et al., 2013)

Because many such phenomena are very sensitive to small variations in climate, especially to temperature, phenological records can be a useful proxy for temperature in historical climatology, especially in the study of climate change and global warming. For example, viticultural records of grape harvests in Europe have been used to reconstruct a record of summer growing season temperatures going back more than 500 years. In addition to providing a longer historical baseline than instrumental measurements, phenological observations provide high temporal resolution of ongoing changes related to global warming.

Past Records

Observations of phenological events have provided indications of the progress of the natural calendar since ancient agricultural times. Many cultures have traditional phenological proverbs and sayings which indicate a time for action: "When the sloe tree is white as a sheet, sow your barley whether it be dry or wet" or attempt to forecast future climate: "If oak's before ash, you're in for a splash. If ash before oak, you're in for a soak". But the indications can be pretty unreliable, as an alternative version of the rhyme shows: "If the oak is out before the ash, 'Twill be a summer of wet and splash; If the ash is out before the oak,'Twill be a summer of fire and smoke." Theoretically, though, these are not mutually exclusive, as one forecasts immediate conditions and one forecasts future conditions.

The North American Bird Phenology Program at USGS Patuxent Wildlife Research Center (PWRC) is in possession of a collection of millions of bird arrival and departure date records for over 870 species across North America, dating between 1880 and 1970. This program, originally started by Wells W. Cooke, involved over 3,000 observers including many notable naturalists of the time. The program ran for 90 years and came to a close in 1970 when other programs starting up at PWRC took precedence. The program was again started in 2009 to digitize the collection of records and now with the help of citizens worldwide, each record is being transcribed into a database which will be publicly accessible for use.

The English naturalists Gilbert White and William Markwick reported the seasonal events of more than 400 plant and animal species, Gilbert White in Selborne, Hampshire and William Markwick in Battle, Sussex over a 25-year period between 1768 and 1793. The data, reported in White's *Natural History and Antiquities of Selborne* are reported as the earliest and latest dates for each event over 25 years; so annual changes cannot therefore be determined.

In Japan and China the time of blossoming of cherry and peach trees is associated with ancient festivals and some of these dates can be traced back to the eighth century. Such historical records may, in principle, be capable of providing estimates of climate at dates before instrumental records became available. For example, records of the harvest dates of the pinot noir grape in Burgundy have been used in an attempt to reconstruct spring–summer temperatures from 1370 to 2003; the reconstructed values during 1787–2000 have a correlation with Paris instrumental data of about 0.75.

Modern Records

Robert Marsham is the founding father of modern phenological recording. Marsham was a wealthy landowner who kept systematic records of "Indications of spring" on his estate at Stratton Strawless, Norfolk, from 1736. These were in the form of dates of the first occurrence of events such as flowering, bud burst, emergence or flight of an insect. Consistent records of the same events or "phenophases" were maintained by generations of the same family over unprecedentedly long periods of time, eventually ending with the death of Mary Marsham in 1958, so that trends can be observed and related to long-term climate records. The data show significant variation in dates which broadly correspond with warm and cold years. Between 1850 and 1950 a long-term trend of gradual climate warming is observable, and during this same period the Marsham record of oak leafing dates tended to become earlier.

After 1960 the rate of warming accelerated, and this is mirrored by increasing earliness of oak leafing, recorded in the data collected by Jean Combes in Surrey. Over the past 250 years, the first leafing date of oak appears to have advanced by about 8 days, corresponding to overall warming on the order of 1.5 °C in the same period.

Towards the end of the 19th century the recording of the appearance and development of plants and animals became a national pastime, and between 1891 and 1948 a programme of phenological recording was organised across the British Isles by the Royal Meteorological Society (RMS). Up to 600 observers submitted returns in some years, with numbers averaging a few hundred. During this period 11 main plant phenophases were consistently recorded over the 58 years from 1891–1948, and a further 14 phenophases were recorded for the 20 years between 1929 and 1948. The returns were summarised each year in the Quarterly Journal of the RMS as *The Phenological Reports*. The 58-year data have been summarised by Jeffree (1960), and show that flowering dates could be as many as 21 days early and as many as 34 days late, with extreme earliness greatest in summer flowering species, and extreme lateness in spring flowering species. In all 25 species, the timings of all phenological events are significantly related to temperature, indicating that phenological events are likely to get earlier as climate warms.

The Phenological Reports ended suddenly in 1948 after 58 years, and Britain was without a national recording scheme for almost 50 years, just at a time when climate change was becoming evident. During this period, important contributions were made by individual dedicated observers. The naturalist and author Richard Fitter recorded the First Flowering Date (FFD) of 557 species of British flowering plants in Oxfordshire between about 1954 and 1990. Writing in Science in 2002, Richard Fitter and his son Alistair Fitter found that "the average FFD of 385 British plant species has advanced by 4.5 days during the past decade compared with the previous four decades." They note that FFD is sensitive to temperature, as is generally agreed, that "150 to 200 species may be flowering on average 15 days earlier in Britain now than in the very recent past" and that these earlier FFDs will have "profound ecosystem and evolutionary consequences".

In the last decade, national recording in Britain has been resumed by the UK Phenology network, run by Woodland Trust and the Centre for Ecology and Hydrology and the BBC Springwatch survey. There is a USA National Phenology Network in which both professional scientists and lay recorders participate, a European Phenology Network that has monitoring, research and educational remits and many other countries such as Canada (Alberta Plantwatch and Saskatchewan PlantWatch), China and Australia have phenological programs.

In eastern North America, almanacs are traditionally used for information on action phenology (in agriculture), taking into account the astronomical positions at the time. William Felker has studied phenology in Ohio, USA since 1973 and now publishes "Poor Will's Almanack", a phenological almanac for farmers (not to be confused with a late 18th century almanac by the same name).

Airborne Sensors

Recent technological advances in studying the earth from space have resulted in a new field of phenological research that is concerned with observing the phenology of whole ecosystems and stands of vegetation on a global scale using proxy approaches. These methods complement the traditional phenological methods which recorded the first occurrences of individual species and phenophases.

The most successful of these approaches is based on tracking the temporal change of a Vegetation Index (like Normalized Difference Vegetation Index(NDVI)). NDVI makes use of the vegetation's typical low reflection in the red (red energy is mostly absorbed by growing plants for Photosynthesis) and strong reflection in the Near Infrared (Infrared energy is mostly reflected by plants due to their cellular structure). Due to its robustness and simplicity, NDVI has become one of the most popular remote sensing based products. Typically, a vegetation index is constructed in such a way that the attenuated reflected sunlight energy (1% to 30% of incident sunlight) is amplified by ratioing red and NIR following this equation:

$$NDVI = \frac{NIR - red}{NIR + red}$$

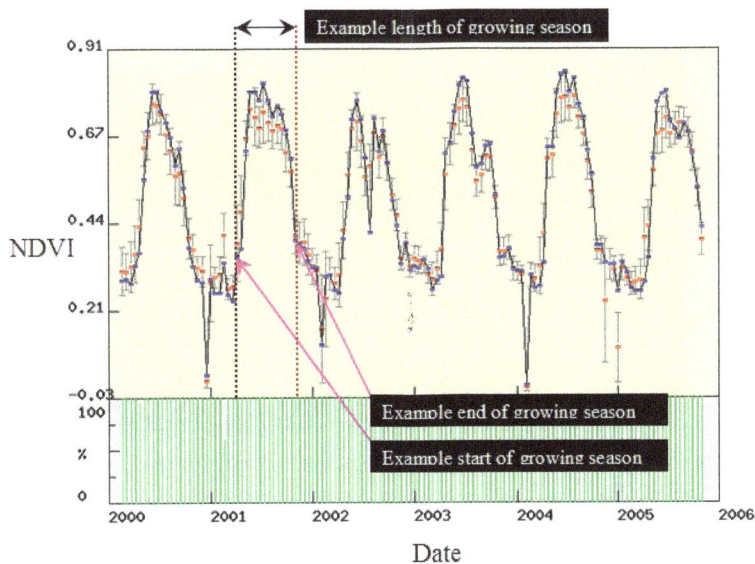

NDVI temporal profile for a typical patch of coniferous forest over a period of six years. This temporal profile depicts the growing season every year as well as changes in this profile from year to year due to climatic and other constraints. Data and graph are based on the MODIS sensor standard public vegetation index product. Data archived at the ORNL DAAC, courtesy of Dr. Robert Cook.

The evolution of the vegetation index through time, depicted by the graph above, exhibits a strong correlation with the typical green vegetation growth stages (emergence, vigor/growth, maturity, and harvest/senescence). These temporal curves are analyzed to extract useful parameters about the vegetation growing season (start of season, end of season, length of growing season, etc.).

Other growing season parameters could potentially be extracted, and global maps of any of these growing season parameters could then be constructed and used in all sorts of climatic change studies.

A noteworthy example of the use of remote sensing based phenology is the work of Ranga Myneni from Boston University. This work showed an apparent increase in vegetation productivity that most likely resulted from the increase in temperature and lengthening of the growing season in the boreal forest. Another example based on the MODIS enhanced vegetation index (EVI) reported by Alfredo Huete at the University of Arizona and colleagues showed that the Amazon Rainforest, as opposed to the long held view of a monotonous growing season or growth only during the wet rainy season, does in fact exhibit growth spurts during the dry season.

However, these phenological parameters are only an approximation of the true biological growth stages. This is mainly due to the limitation of current space-based remote sensing, especially the spatial resolution, and the nature of vegetation index. A pixel in an image does not contain a pure target (like a tree, a shrub, etc.) but contains a mixture of whatever intersected the sensor's field of view.

Nutrient Management

Nitrogen fertilizer being applied to growing corn (maize) in a contoured, no-tilled field in Iowa.

Nutrient management is the science and art directed to link soil, crop, weather, and hydrologic factors with cultural, irrigation, and soil and water conservation practices to achieve the goals of optimizing nutrient use efficiency, yields, crop quality, and economic returns, while reducing off-site transport of nutrients that may impact the environment. Nutrient management is the skillful task of matching a specific field soil, climate, and crop management conditions to rate, source, timing, and place (commonly known as the 4R nutrient stewardship) of nutrient application.

Some important factors that need to be considered when managing nutrients include (a) the application of nutrients considering the achievable optimum yields and, in some cases, crop quality; (b) the management, application, and timing of nutrients using a budget based on all sources and sinks active at the site; and (c) the management of soil, water, and crop to minimize the off-site transport of nutrients from nutrient leaching out of the root zone, surface runoff, and volatilization (or other gas exchanges).

There can be potential interactions because of differences in nutrient pathways and dynamics. For instance, practices that reduce the off-site surface transport of a given nutrient may increase the leaching losses of other nutrients. These complex dynamics present nutrient managers the difficult task of integrating soil, crop, weather, hydrology, and management practices to achieve the best balance for maximizing profit while contributing to the conservation of our biosphere.

Nutrient Management Plan

Manure spreader

A crop nutrient management plan is a tool that farmers can use to increase the efficiency of all the nutrient sources a crop uses while reducing production and environmental risk, ultimately increasing profit. It is generally agreed that there are ten fundamental components of a Crop Nutrient Management Plan. Each component is critical to helping analyze each field and improve nutrient efficiency for the crops grown. These components include:

Field Map: The map, including general reference points (such as streams, residences, wellheads etc.), number of acres, and soil types is the base for the rest of the plan.

Soil Test: How much of each nutrient (N-P-K and other critical elements such as pH and organic matter) is in the soil profile? The soil test is a key component needed for developing the nutrient rate recommendation.

Crop Sequence: Did the crop that grew in the field last year (and in many cases two or more years ago) fix nitrogen for use in the following years? Has long-term no-till increased organic matter? Did the end-of-season stalk test show a nutrient deficiency? These factors also need to be factored into your plan.

Estimated Yield: Factors that affect yield are numerous and complex. A field's soils, drainage, insect, weed and disease pressure, rotation and many other factors differentiate one field from another. This is why using historic yields is important in developing yield estimates for next year. Accurate yield estimates can dramatically improve nutrient use efficiency.

Sources and Forms: The sources and forms of available nutrients can vary from farm-to-farm and even field-to-field. For instance, manure fertility analysis, storage practices and other factors will need to be included in a nutrient management plan. Manure nutrient tests/analysis are one way to determine the fertility of it. Nitrogen fixed from a previous year's legume crop and residual affects of manure also effects rate recommendations. Many other nutrient sources should also be factored into this plan.

Sensitive Areas: What's out of the ordinary about a field's plan? Is it irrigated? Next to a stream or lake? Especially sandy in one area? Steep slope or low area? Manure applied in one area for generations due to proximity of dairy barn? Extremely productive—or unproductive—in a portion of the field? Are there buffers that protect streams, drainage ditches, wellheads, and other water collection points? How far away are the neighbors? What's the general wind direction? This is the place to note these and other special conditions that need to be considered.

Recommended rates: Here's the place where science, technology, and art meet. Given everything you've noted, what is the optimum rate of N, P, K, lime and any other nutrients? While science tells us that a crop has changing nutrient requirements during the growing season, a combination of technology and farmer's management skills assure optimum nutrient availability at all stages of growth. No-till corn generally requires starter fertilizer to give the seedling a healthy start.

Recommended timing: When does the soil temperature drop below 50 degrees? Will a N stabilizer be used? What's the tillage practice? Strip-till corn and no-till often require different timing approaches than seed planted into a field that's been tilled once with a field cultivator. Will a starter fertilizer be used to give the seedling a healthy start? How many acres can be covered with available labor (custom or hired) and equipment? Does manure application in a farm depend on a custom applicator's schedule? What agreements have been worked out with neighbors for manure use on their fields? Is a neighbor hosting a special event? All these factors and more will likely figure into the recommended timing.

Recommended methods: Surface or injected? While injection is clearly preferred, there may be situations where injection is not feasible (i.e. pasture, grassland). Slope, rainfall patterns, soil type, crop rotation and many other factors determine which method is best for optimizing nutrient efficiency (availability and loss) in farms. The combination that's right in one field may differ in another field even with the same crop.

Annual review and update: Even the best managers are forced to deviate from their plans. What rate was actually applied? Where? Using which method? Did an unusually mild winter or wet spring reduce soil nitrate? Did a dry summer, disease, or some other unusual factor increase nutrient carryover? These and other factors should be noted. It's easier to make notes throughout the year than to remember back six to 10 months.

When such a plan is designed for animal feeding operations (AFO), it may be termed a "manure management plan." In the United States, some regulatory agencies recommend or require that farms implement these plans in order to prevent water pollution. The U.S. Natural Resources Conservation Service (NRCS) has published guidance documents on preparing a comprehensive nutrient management plan (CNMP) for AFOs.

The International Plant Nutrition Institute has published a 4R plant nutrition manual for improving the management of plant nutrition. The manual outlines the scientific principles behind each of the four R's or "rights" and discusses the adoption of 4R practices on the farm, approaches to nutrient management planning, and measurement of sustainability performance.

Food Systems

The term "food system" is used frequently in discussions about nutrition, food, health, community economic development and agriculture. A food system includes all processes and infrastructure involved in feeding a population: growing, harvesting, processing, packaging, transporting, marketing, consumption, and disposal of food and food-related items. It also includes the inputs needed and outputs generated at each of these steps. A food system operates within and is influenced by social, political, economic and environmental contexts. It also requires human resources that provide labor, research and education. Food systems are either conventional or alternative according to their model of food lifespan from origin to plate.

Conventional Food Systems

Conventional food systems operate on the economies of scale. These food systems are geared towards a production model that requires maximizing efficiency in order to lower consumer costs and increase overall production, and they utilize economic models such as vertical integration, economic specialization, and global trade.

The term "conventional" when describing food systems is large part due to comparisons made to it by proponents of other food systems, collectively known as alternative food systems.

History of Conventional Food Systems

The development of food systems can be traced back to the origins of in-situ agriculture and the production of food surpluses. These surpluses enabled the development of settled areas and contributed to the development of ancient civilizations, particularly those in the Fertile Crescent. The system of trade associated with the exchange of foodstuffs also emerged in East Asia, North America, South America, and Subsaharan Africa with common commodities of exchange such as salt, spices, fish, grains, etc. Through events in world history such as the conquests of Alexander the Great, the Crusades, the expansion of Islam, the journeys of Marco Polo, and the exploration and colonization of the Americas by Europeans led to the introduction and redistribution of new foods to the world at large, and food systems began to intermingle on a global scale. After World War II, the advent of industrialized agriculture and more robust global trade mechanisms have evolved into the models of food production, presentation, delivery, and disposal that characterizes conventional food systems today.

Impacts of Conventional Food Systems

Lower food costs and greater food variety can be directly attributed to the evolvement of conventional food systems. Agronomic efficiency is driven by the necessity to constantly lower production expenses, and those savings can then be passed on to the consumer. Also, the advent of industrial agriculture and the infrastructure built around conventional food systems has enabled the world population to expand beyond the "Malthusian catastrophe" limitations.

However, conventional food systems are largely based on the availability of inexpensive fossil fuels, which is necessary for mechanized agriculture, the manufacture or collection of chemical fertilizers, the processing of food products, and the packaging of the foods. Food processing began when the number of consumers started growing rapidly. The demand for cheap and efficient calories climbed resulting in nutrition decline. Industrialized agriculture, due to its reliance on economies of scale to reduce production costs, often leads to the compromising of local, regional, or even global ecosystems through fertilizer runoff, nonpoint source pollution, and greenhouse gas emission. Also, the need to reduce production costs in an increasingly global market can cause production of foods to be moved to areas where economic costs (labor, taxes, etc.) are lower or environmental regulations are more lax, which are usually further from consumer markets. For example, the majority of salmon sold in the United States is raised off the coast of Chile, due in large part to less stringent Chilean standards regarding fish feed and regardless of the fact that salmon are not indigenous in Chilean coastal waters. The globalization of food production can result in the loss of traditional food systems in less developed countries, and have negative impacts on the population health, ecosystems, and cultures in those countries.

Alternative Food Systems

Alternative food systems are those that fall outside the scope of conventional agriculture.

Local Food Systems

Local food systems are networks of food production and consumption that aim to be geographically and economically accessible and direct. They contrast to industrial food systems by operating with reduced food transportation and more direct marketing, leading to fewer people between the farmer and the consumer. As a result, relationships that are developed in local food systems emerge from face-to-face interactions, potentially leading to a stronger sense of trust and social connectedness between actors. As a result, some scholars suggest that local food systems are a good way to revitalize a community. The decreased distance of food transportation has also been promoted for its environmental benefits.

Both proponents and critics of local food systems warn that they can lead to narrow inward-looking attitudes or 'local food patriotism', and that price premiums and local food cultures can be elitist and exclusive.

Examples of local food systems include community-supported agriculture, farmers markets and farm to school programs. They have been associated with the 100 Mile Diet and Low Carbon Diet, as well as the food sovereignty movement and slow food movement. Various forms of urban agriculture locate food production in densely populated areas not traditionally associated with farming. Garden sharing, where urban and suburban homeowners offer land access to food growers in exchange for a share of the harvest, is a relatively new trend, at the extreme end of direct local food production.

Organic Food Systems

Organic food systems are characterized by a reduced dependence on chemical inputs and an increased concern for transparency and information. Organic produce is grown without the chemical pesticides and fertilizers of industrial food systems, and livestock is reared without the use of antibiotics or growth hormones. The reduced inputs of organic agriculture can also lead to a greater reliance on local knowledge, creating a stronger knowledge community amongst farmers. The transparency of food information is vital for organic food systems as a means through which consumers are able to identify organic food. As a result, a variety of certification bodies have emerged in organic food systems that set the standards for organic identification. Organic agriculture is promoted for the ecological benefits of reduced chemical application, the health benefits of lower chemical consumption, the economic benefits that accrue to farmers through a price premium, and the social benefits of increased transparency in the food system.

Like local food systems, organic food systems have been criticized for being elitist and inaccessible. Critics have also suggested that organic agriculture has been conventionalized such that it mimics industrial food systems while using pesticides and fertilizers that are organically derived

Cooperatives in Food Systems

Cooperatives can exist both at the farmer end of food production and the consumer end. Farming cooperatives refer to arrangements where farmers pool resources, either to cultivate their crops or get their crops to market. Consumer cooperatives often refer to food cooperatives where members buy a share in the store. Co-operative grocery stores, unlike corporate grocery stores, are socially owned and thus surpluses cannot be taken from the store as profit. As a result, food co-ops do not work for profit, potentially keeping prices more cost representative. Other forms of cooperatives that have developed more recently include community-supported agriculture, where community members buy a share in a farm's harvest, and may also be engaged in farm labour, operating at both the consumer and producer end of food systems. Garden sharing pairs individual landowners and food growers, while variations on this approach organize groups of food gardeners for mutual assistance.

The benefits of cooperatives are largely in the redistribution of risk and responsibility. For farming cooperatives that share resources, the burden of investment is disbursed to all members, rather than being concentrated in a single individual. A criticism of cooperatives is that reduced competition can reduce efficiency

Alternative Food for Global Catastrophic Risk

A global catastrophic risk is a hypothetical future event with the potential to inflict serious damage to human well-being on a global scale. Some such events could destroy or cripple modern civilization. These global catastrophes include: super volcanic eruption, asteroid or comet impact, nuclear winter, abrupt climate change, super weed, super crop pathogen, super bacterium, or super crop pest. Mass human starvation is currently likely if global agricultural production is dramatically reduced for several years following any such event. Engineers, David Denkenberger and Joshua Pearce argue in *Feeding Everyone No Matter What* that humanity could be fed in such situations by converting natural gas and wood fiber to human edible food using known pathways. Potential pathways include growing mushrooms on wood and then eating the mushrooms directly as well as feeding the waste to ruminates and then eating them.

Fair Trade

Fair trade has emerged in global food systems to create a greater balance between the price of food and the cost of producing it. It is defined largely by more direct trading systems whereby producers have greater control over the conditions of trade and garner a greater fraction of the sale price. The main goal of Fair Trade is to "change international commercial relations in such a way that disadvantaged producers can increase their control over their own future, have a fair and just return for their work, continuity of income and decent working and living conditions through sustainable development" Like organic food systems, fair trade relies on transparency and the flow of information. Well-known examples of fair trade commodities are coffee and cocoa.

Transparency

Transparency within food systems refers to full disclosure of information about rules, procedures and practices at all levels within a food production and supply chain. Transparency ensures that consumers have detailed information about production of a given food item. Traceability, by contrast, is the ability to trace to their origins all components in a food production and marketing chain, whether processed or unprocessed (e.g., meat, vegetables) foods.2 Concerns about transparency and traceability have been heightened with food safety scares such as Bovine Spongiform Encephalopathy (BSE) and Escherichia coli (E. coli), but do not exclusively refer to food safety. Transparency is also important in identifying foods that possess extrinsic qualities that do not affect the nature of the food per se, but affect its production, such as animal welfare, social justice issues and environmental concerns.

One of the primary ways transparency is achieved is through certification and/or use of food labels. In the United States, some certification originates in the public sector, such as the United States Department of Agriculture (USDA) Organic label. Others have their origin in private sector certification (e.g., Humanely Raised, Certified Humane). There are also labels which do not rely on certification, such as the USDA's Country of Origin Label (COOL).

Participation in local food systems such as Community Supported Agriculture (CSA), Farmers Markets, food cooperatives and farmer cooperatives also enhances transparency, and there are diverse programs promoting purchase of locally grown and marketed foods.

Labeling

USDA ORGANIC USDA Organic Label	Organic (USA) - The USDA Organic label indicates that the product has been produced in accordance with the USDA's Federal Organic Standard. This label is applied to fruits, vegetables, meat, eggs and dairy products. Some states, such as California, have their own organic label. Organic labelling is prominent internationally as well.

 Fair Trade Show in UK	Fair Trade -Indicates that the product has been grown and marketed in accordance with Fair Trade standards. This is an independent certification, awarded by FLO-CERT and overseen by FLO International. Major food items that are marketed under Fair Trade are coffee, tea and chocolate. Many items other than food are sold with a Fair Trade label.
	Food Alliance Certified. Food Alliance is a nonprofit organization that certifies farms, ranches, and food processors and distributors for safe and fair working conditions, humane treatment of animals, and good environmental stewardship. Food Alliance Certified products come from farms, ranches and food processors that have met meaningful standards for social and environmental responsibility, as determined through an independent third-party audit. Food Alliance does not certify genetically modified crops or livestock. Meat or dairy products come from animals that are not treated with antibiotics or growth hormones. Food Alliance Certified foods never contain artificial colors, flavors, or preservatives. Food Alliance Certified. http://www.foodalliance.org
 Examples of COOL Labeling	Country of Origin - This label was created by enactment of the 2002 Farm Bill. The US Department of Agriculture is responsible for its implementation, which began 30 September 2008. The bill mandates country of origin labeling for several products, including beef, lamb, pork, fish, chicken, perishable agricultural commodities and some nuts. USDA rules provide specifics as to documentation, timetables and definitions. There is not one specific label to indicate the country of origin; they will vary by country.
	American Humane Certified. This certification is provided by the American Humane Association, and ensures that farm animals are raised according to welfare standards that provide for adequate housing, feed, healthcare and behavior expression. Antibiotics are not used except for therapeutic reasons; growth promoters are not used. Other issues including transport, processing and biosecurity are addressed as well. Species covered are poultry, cattle and swine.
	Certified Humane Raised & Handled. This label ensures that production meets the Humane Farm Animal Care Program standards, which addresses housing, diet (excluding routine use of hormones or antibiotics) and natural behavior. Additionally, producers must comply with food safety and environmental protection regulations. They must meet standards set by the American Meat Institute, that are more stringent than those laid out in the Federal Humane Slaughter Act. Certification has been applied to beef, poultry and eggs, pork, lamb, goat, turkey, veal, dairy products and wool.

References

- Encyclopedia of Agrophysics in series: Encyclopedia of Earth Sciences Series edts. Jan Glinski, Jozef Horabik, Jerzy Lipiec, 2011, Publisher: Springer, ISBN 978-90-481-3585-1

- Scientific Dictionary of Agrophysics: polish-English, polsko-angielski by R. Dębicki, J. Gliński, J. Horabik, R. T. Walczak - Lublin 2004, ISBN 83-87385-88-3

- Physical Methods in Agriculture. Approach to Precision and Quality, edts. J. Blahovec and M. Kutilek, Kluwer Academic Publishers, New York 2002, ISBN 0-306-47430-1.

- Soil Aeration and its Role for Plants by J. Gliński, W. Stępniewski, 1985, Publisher: CRC Press, Inc., Boca Raton, USA, ISBN 0-8493-5250-9

- Nestle, Marion. (2013). Food Politics: How the Food Industry Influences Nutrition and Health." Los Angeles, California: University of California Press. ISBN 978-0520275966

- Denkenberger, D. C., & Pearce, J. M. (2015). Feeding Everyone: Solving the Food Crisis in Event of Global Catastrophes that Kill Crops or Obscure the Sun. Futures. 72:57–68. open access

Techniques and Practices of Agronomy

Agronomy focuses on efficient techniques of cultivation, pest control and plant breeding. It also aims to aims to protect the soil through the cultivation of cover crops. These practices change according to region as the themes like shifting cultivation and crop rotation will show. Organic farming is a major branch of agronomic practice.

Cover Crop

A cover crop is a crop planted primarily to manage soil erosion, soil fertility, soil quality, water, weeds, pests, diseases, biodiversity and wildlife in an *agroecosystem* (Lu *et al.* 2000), an ecological system managed and largely shaped by humans across a range of intensities to produce food, feed, or fiber. Currently, not many countries are known for using the cover crop method.

Cover crops are of interest in sustainable agriculture as many of them improve the sustainability of agroecosystem attributes and may also indirectly improve qualities of neighboring natural ecosystems. Farmers choose to grow and manage specific cover crop types based on their own needs and goals, influenced by the biological, environmental, social, cultural, and economic factors of the food system in which they operate (Snapp *et al.* 2005). The farming practice of cover crops has been recognized as climate-smart agriculture by the White House.

Soil Erosion

Although cover crops can perform multiple functions in an agroecosystem simultaneously, they are often grown for the sole purpose of preventing soil erosion. Soil erosion is a process that can irreparably reduce the productive capacity of an agroecosystem. Dense cover crop stands physically slow down the velocity of rainfall before it contacts the soil surface, preventing soil splashing and erosive surface runoff (Romkens *et al.* 1990). Additionally, vast cover crop root networks help anchor the soil in place and increase soil porosity, creating suitable habitat networks for soil macrofauna (Tomlin *et al.* 1995).

Soil Fertility Management

One of the primary uses of cover crops is to increase soil fertility. These types of cover crops are referred to as "green manure." They are used to manage a range of soil macronutrients and micronutrients. Of the various nutrients, the impact that cover crops have on nitrogen management has received the most attention from researchers and farmers, because nitrogen is often the most limiting nutrient in crop production.

Often, green manure crops are grown for a specific period, and then plowed under before reaching full maturity in order to improve soil fertility and quality. Also the stalks left block the soil from being eroded.

Green manure crops are commonly leguminous, meaning they are part of the Fabaceae (pea) family.This family is unique in that all of the species in it set pods, such as bean, lentil, lupins and alfalfa. Leguminous cover crops are typically high in nitrogen and can often provide the required quantity of nitrogen for crop production. In conventional farming, this nitrogen is typically applied in chemical fertilizer form. This quality of cover crops is called fertilizer replacement value (Thiessen-Martens *et al.* 2005).

Another quality unique to leguminous cover crops is that they form symbiotic relationships with the rhizobial bacteria that reside in legume root nodules. Lupins is nodulated by the soil microorganism *Bradyrhizobium* sp. (Lupinus). Bradyrhizobia are encountered as microsymbionts in other leguminous crops (*Argyrolobium, Lotus, Ornithopus, Acacia, Lupinus*) of Mediterranean origin. These bacteria convert biologically unavailable atmospheric nitrogen gas (N2) to biologically available ammonium (NH+ 4) through the process of biological nitrogen fixation.

Prior to the advent of the Haber-Bosch process, an energy-intensive method developed to carry out industrial nitrogen fixation and create chemical nitrogen fertilizer, most nitrogen introduced to ecosystems arose through biological nitrogen fixation (Galloway *et al.* 1995). Some scientists believe that widespread biological nitrogen fixation, achieved mainly through the use of cover crops, is the only alternative to industrial nitrogen fixation in the effort to maintain or increase future food production levels (Bohlool *et al.* 1992, Peoples and Craswell 1992, Giller and Cadisch 1995). Industrial nitrogen fixation has been criticized as an unsustainable source of nitrogen for food production due to its reliance on fossil fuel energy and the environmental impacts associated with chemical nitrogen fertilizer use in agriculture (Jensen and Hauggaard-Nielsen 2003). Such widespread environmental impacts include nitrogen fertilizer losses into waterways, which can lead to eutrophication (nutrient loading) and ensuing hypoxia (oxygen depletion) of large bodies of water.

An example of this lies in the Mississippi Valley Basin, where years of fertilizer nitrogen loading into the watershed from agricultural production have resulted in a hypoxic "dead zone" off the Gulf of Mexico the size of New Jersey (Rabalais *et al.* 2002). The ecological complexity of marine life in this zone has been diminishing as a consequence (CENR 2000).

As well as bringing nitrogen into agroecosystems through biological nitrogen fixation, types of cover crops known as "catch crops" are used to retain and recycle soil nitrogen already present. The catch crops take up surplus nitrogen remaining from fertilization of the previous crop, preventing it from being lost through leaching (Morgan *et al.* 1942), or gaseous denitrification or volatilization (Thorup-Kristensen *et al.* 2003).

Catch crops are typically fast-growing annual cereal species adapted to scavenge available nitrogen efficiently from the soil (Ditsch and Alley 1991). The nitrogen tied up in catch crop biomass is released back into the soil once the catch crop is incorporated as a green manure or otherwise begins to decompose.

An example of green manure use comes from Nigeria, where the cover crop *Mucuna pruriens* (velvet bean) has been found to increase the availability of phosphorus in soil after a farmer applies rock phosphate (Vanlauwe *et al.* 2000).

Soil Quality Management

Cover crops can also improve soil quality by increasing soil organic matter levels through the input of cover crop biomass over time. Increased soil organic matter enhances soil structure, as well as the water and nutrient holding and buffering capacity of soil (Patrick *et al.* 1957). It can also lead to increased soil carbon sequestration, which has been promoted as a strategy to help offset the rise in atmospheric carbon dioxide levels (Kuo *et al.* 1997, Sainju *et al.* 2002, Lal 2003).

Soil quality is managed to produce optimum circumstances for crops to flourish. The principal factors of soil quality are soil salination, pH, microorganism balance and the prevention of soil contamination.

Water Management

By reducing soil erosion, cover crops often also reduce both the rate and quantity of water that drains off the field, which would normally pose environmental risks to waterways and ecosystems downstream (Dabney *et al.* 2001). Cover crop biomass acts as a physical barrier between rainfall and the soil surface, allowing raindrops to steadily trickle down through the soil profile. Also, as stated above, cover crop root growth results in the formation of soil pores, which in addition to enhancing soil macrofauna habitat provides pathways for water to filter through the soil profile rather than draining off the field as surface flow. With increased water infiltration, the potential for soil water storage and the recharging of aquifers can be improved (Joyce *et al.* 2002).

Just before cover crops are killed (by such practices including mowing, tilling, discing, rolling, or herbicide application) they contain a large amount of moisture. When the cover crop is incorporated into the soil, or left on the soil surface, it often increases soil moisture. In agroecosystems where water for crop production is in short supply, cover crops can be used as a mulch to conserve water by shading and cooling the soil surface. This reduces evaporation of soil moisture. In other situations farmers try to dry the soil out as quickly as possible going into the planting season. Here prolonged soil moisture conservation can be problematic.

While cover crops can help to conserve water, in temperate regions (particularly in years with below average precipitation) they can draw down soil water supply in the spring, particularly if climatic growing conditions are good. In these cases, just before crop planting, farmers often face a tradeoff between the benefits of increased cover crop growth and the drawbacks of reduced soil moisture for cash crop production that season. C/N ratio is balanced with this application.

Weed Management

Thick cover crop stands often compete well with weeds during the cover crop growth period, and can prevent most germinated weed seeds from completing their life cycle and reproducing. If the cover crop is left on the soil surface rather than incorporated into the soil as a green manure after its growth is terminated, it can form a nearly impenetrable mat. This drastically reduces light transmittance to weed seeds, which in many cases reduces weed seed germination rates (Teasdale 1993). Furthermore, even when weed seeds germinate, they often run out of stored energy for growth before building the necessary structural capacity to break through the cover crop mulch layer. This is often termed the *cover crop smother effect* (Kobayashi *et al.* 2003).

Cover crop in South Dakota

Some cover crops suppress weeds both during growth and after death (Blackshaw *et al.* 2001). During growth these cover crops compete vigorously with weeds for available space, light, and nutrients, and after death they smother the next flush of weeds by forming a mulch layer on the soil surface. For example, Blackshaw *et al.* (2001) found that when using *Melilotus officinalis* (yellow sweetclover) as a cover crop in an improved fallow system (where a fallow period is intentionally improved by any number of different management practices, including the planting of cover crops), weed biomass only constituted between 1-12% of total standing biomass at the end of the cover crop growing season. Furthermore, after cover crop termination, the yellow sweetclover residues suppressed weeds to levels 75-97% lower than in fallow (no yellow sweetclover) systems.

In addition to competition-based or physical weed suppression, certain cover crops are known to suppress weeds through allelopathy (Creamer *et al.* 1996, Singh *et al.* 2003). This occurs when certain biochemical cover crop compounds are degraded that happen to be toxic to, or inhibit seed germination of, other plant species. Some well known examples of allelopathic cover crops are *Secale cereale* (rye), *Vicia villosa* (hairy vetch), *Trifolium pratense* (red clover), *Sorghum bicolor* (sorghum-sudangrass), and species in the Brassicaceae family, particularly mustards (Haramoto and Gallandt 2004). In one study, rye cover crop residues were found to have provided between 80% and 95% control of early season broadleaf weeds when used as a mulch during the production of different cash crops such as soybean, tobacco, corn, and sunflower (Nagabhushana *et al.* 2001).

In a recent study released by the Agricultural Research Service (ARS) scientists examined how rye seeding rates and planting patterns affected cover crop production. The results show that planting more pounds per acre of rye increased the cover crop's production as well as decreased the amount of weeds. The same was true when scientists tested seeding rates on legumes and oats; a higher density of seeds planted per acre decreased the amount of weeds and increased the yield of legume and oat production. The planting patterns, which consisted of either traditional rows or grid patterns, did not seem to make a significant impact on the cover crop's production or on the weed production in either cover crop. The ARS scientists concluded that increased seeding rates could be an effective method of weed control.

Disease Management

In the same way that allelopathic properties of cover crops can suppress weeds, they can also break disease cycles and reduce populations of bacterial and fungal diseases (Everts 2002), and parasitic nematodes (Potter *et al.* 1998, Vargas-Ayala *et al.* 2000). Species in the Brassicaceae family, such as mustards, have been widely shown to suppress fungal disease populations through the release of naturally occurring toxic chemicals during the degradation of glucosinolade compounds in their plant cell tissues (Lazzeri and Manici 2001).

Pest Management

Some cover crops are used as so-called "trap crops", to attract pests away from the crop of value and toward what the pest sees as a more favorable habitat (Shelton and Badenes-Perez 2006). Trap crop areas can be established within crops, within farms, or within landscapes. In many cases the trap crop is grown during the same season as the food crop being produced. The limited area occupied by these trap crops can be treated with a pesticide once pests are drawn to the trap in large enough numbers to reduce the pest populations. In some organic systems, farmers drive over the trap crop with a large vacuum-based implement to physically pull the pests off the plants and out of the field (Kuepper and Thomas 2002). This system has been recommended for use to help control the lygus bugs in organic strawberry production (Zalom *et al.* 2001). Another example of trap crops are nematode resistance White mustard (*Sinapis alba*) and Radish (*Raphanus sativus*). They can be grown after a main (cereal) crop and trap nematodes, for example the beet cyst nematode and Columbian root knot nematode. When grown, nematodes hatch and are attracted to the roots. After entering the roots they cannot reproduce in the root due to a hypersensitive resistance reaction of the plant. Hence the nematode population is greatly reduced, by 70-99%, depending on species and cultivation time.

Other cover crops are used to attract natural predators of pests by providing elements of their habitat. This is a form of biological control known as habitat augmentation, but achieved with the use of cover crops (Bugg and Waddington 1994). Findings on the relationship between cover crop presence and predator/pest population dynamics have been mixed, pointing toward the need for detailed information on specific cover crop types and management practices to best complement a given integrated pest management strategy. For example, the predator mite *Euseius tularensis* (Congdon) is known to help control the pest citrus thrips in Central California citrus orchards. Researchers found that the planting of several different leguminous cover crops (such as bell bean, woollypod vetch, New Zealand white clover, and Austrian winter pea) provided sufficient pollen as a feeding source to cause a seasonal increase in *E. tularensis* populations, which with good timing could potentially introduce enough predatory pressure to reduce pest populations of citrus thrips (Grafton-Cardwell *et al.* 1999).

Diversity and Wildlife

Although cover crops are normally used to serve one of the above discussed purposes, they often simultaneously improve farm habitat for wildlife. The use of cover crops adds at least one more dimension of plant diversity to a cash crop rotation. Since the cover crop is typically not a crop of value, its management is usually less intensive, providing a window of "soft" human influence on the farm. This relatively "hands-off" management, combined with the increased on-farm heterogeneity created by the establishment of cover crops, increases the likelihood that a more complex trophic structure will develop to support a higher level of wildlife diversity (Freemark and Kirk 2001).

In one study, researchers compared arthropod and songbird species composition and field use between conventionally and cover cropped cotton fields in the Southern United States. The cover cropped cotton fields were planted to clover, which was left to grow in between cotton rows throughout the early cotton growing season (stripcover cropping). During the migration and breeding season, they found that songbird densities were 7–20 times higher in the cotton fields with integrated clover cover crop than in the conventional cotton fields. Arthropod abundance and biomass was also higher in the clover cover cropped fields throughout much of the songbird breeding season, which was attributed to an increased supply of flower nectar from the clover. The clover cover crop enhanced songbird habitat by providing cover and nesting sites, and an increased food source from higher arthropod populations (Cederbaum *et al.* 2004).

Companion Planting

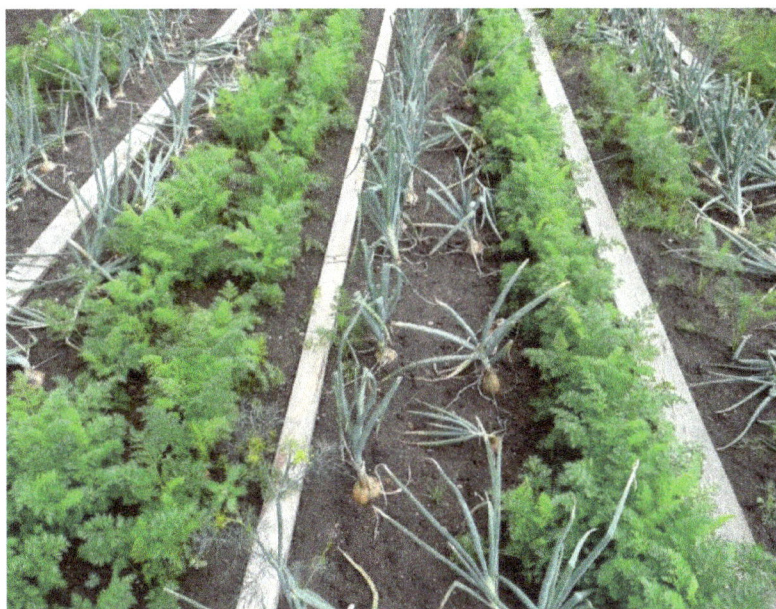

Companion planting of carrots and onions

Companion planting in gardening and agriculture is the planting of different crops in proximity for pest control, pollination, providing habitat for beneficial creatures, maximizing use of space, and to otherwise increase crop productivity. Companion planting is a form of polyculture.

Companion planting is used by farmers and gardeners in both industrialized and developing countries for many reasons. Many of the modern principles of companion planting were present many centuries ago in cottage gardens in England and forest gardens in Asia, and thousands of years ago in Mesoamerica.

History

In China, mosquito ferns (*Azolla* spp.) have been used for at least a thousand years as companion plants for rice crops. They host a cyanobacterium that fixes nitrogen from the atmosphere, and they block light from plants that would compete with the rice.

Companion planting was practiced in various forms by the indigenous peoples of the Americas prior to the arrival of Europeans. These peoples domesticated squash 8,000 to 10,000 years ago, then maize, then common beans, forming the Three Sisters agricultural technique. The cornstalk served as a trellis for the beans to climb, and the beans fixed nitrogen, benefitting the maize.

Companion planting was widely promoted in the 1970s as part of the organic gardening movement. It was encouraged for pragmatic reasons, such as natural trellising, but mainly with the idea that different species of plant may thrive more when close together. It is also a technique frequently used in permaculture, together with mulching, polyculture, and changing of crops.

Examples of Companion Plants

Nasturtium (*Tropaeolum majus*) is a food plant of some caterpillars which feed primarily on members of the cabbage family (brassicas), and some gardeners claim that planting them around brassicas protects the food crops from damage, as eggs of the pests are preferentially laid on the nasturtium. This practice is called trap cropping.

The smell of the foliage of marigolds is claimed to deter aphids from feeding on neighbouring crops. Marigolds with simple flowers also attract nectar-feeding adult hoverflies, the larvae of which are predators of aphids.

Various legume crops benefit from being commingled with a grassy nurse crop. For example, common vetch or hairy vetch is planted together with rye or winter wheat to make a good cover crop or green manure (or both).

The terms "undersowing" and "overseeding" both involve intercropping as a type of companion planting. "Undersowing" conveys the idea of sowing the second crop among the young plants of the first crop (or in between the rows, if rows are used). A connotation of understory growth is conveyed, albeit exaggerated (because the first crop is not yet a dense canopy). "Overseeding" conveys the idea of broadcasting the seeds of the second crop over the existing first crop. This is analogous to overseeding a lawn to improve the mix of grasses present.

Versions

There are a number of systems and ideas using companion planting.

Square foot gardening attempts to protect plants from many normal gardening problems by packing them as closely together as possible, which is facilitated by using companion plants, which can be closer together than normal.

Another system using companion planting is the forest garden, where companion plants are intermingled to create an actual ecosystem, emulating the interaction of up to seven levels of plants in a forest or woodland.

Organic gardening may make use of companion planting, since many synthetic means of fertilizing, weed reduction and pest control are forbidden.

Host-finding Disruption

Recent studies on host-plant finding have shown that flying pests are far less successful if their host-plants are surrounded by any other plant or even "decoy-plants" made of green plastic, cardboard, or any other green material.

The host-plant finding process occurs in phases:

- The first phase is stimulation by odours characteristic to the host-plant. This induces the insect to try to land on the plant it seeks. But insects avoid landing on brown (bare) soil. So if only the host-plant is present, the insects will quasi-systematically find it by simply landing on the only green thing around. This is called (from the point of view of the insect) "appropriate landing". When it does an "inappropriate landing", it flies off to any other nearby patch of green. It eventually leaves the area if there are too many 'inappropriate' landings.

- The second phase of host-plant finding is for the insect to make short flights from leaf to leaf to assess the plant's overall suitability. The number of leaf-to-leaf flights varies according to the insect species and to the host-plant stimulus received from each leaf. The insect must accumulate sufficient stimuli from the host-plant to lay eggs; so it must make a certain number of consecutive 'appropriate' landings. Hence if it makes an 'inappropriate landing', the assessment of that plant is negative, and the insect must start the process anew.

Thus it was shown that clover used as a ground cover had the same disruptive effect on eight pest species from four different insect orders. An experiment showed that 36% of cabbage root flies laid eggs beside cabbages growing in bare soil (which resulted in no crop), compared to only 7% beside cabbages growing in clover (which allowed a good crop). Simple decoys made of green cardboard also disrupted appropriate landings just as well as did the live ground cover.

Companion Plant Categories

The use of companion planting can be of benefit to the grower in a number of different ways, including:

- Hedged investment – the growing of different crops in the same space increases the odds of some yield being given, even if one crop fails.

- Increased level interaction – when crops are grown on different levels in the same space, such as providing ground cover or one crop working as a trellis for another, the overall yield of a plot may be increased.

- Protective shelter is when one type of plant may serve as a wind break or provide shade for another.

- Pest suppression – some companion plants may help prevent pest insects or pathogenic fungi from damaging the crop, through chemical means.

- Predator recruitment and positive hosting – The use of companion plants that produce copious nectar or pollen in a vegetable garden (insectary plants) may help encourage higher populations of beneficial insects that control pests, as some beneficial predatory insects only consume pests in their larval form and are nectar or pollen feeders in their adult form.

- Trap cropping – some companion plants are claimed to attract pests away from others.

- Pattern disruption – in a monoculture pests spread easily from one crop plant to the next, whereas such easy progress may be disrupted by surrounding companion plants of a different type.

Crop Rotation

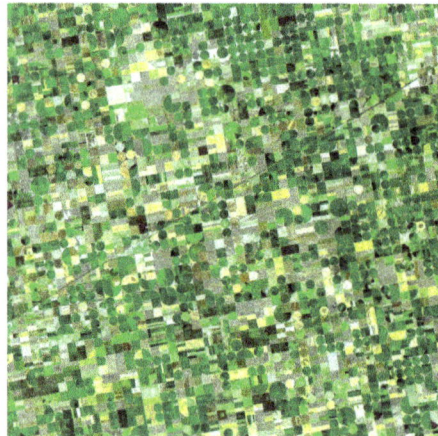

Satellite image of circular crop fields in Kansas in late June 2001. Healthy, growing crops
are green. Corn would be growing into leafy stalks by then. Sorghum, which resembles corn, grows
more slowly and would be much smaller and therefore, (possibly) paler. Wheat is a brilliant yellow as
harvest occurs in June. Fields of brown have been recently harvested and plowed under or lie fallow for the year.

Effects of crop rotation and monoculture at the Swojec Experimental Farm, Wroclaw
University of Environmental and Life Sciences. In the front field, the "Norfolk" crop rotation sequence
(potatoes, oats, peas, rye) is being applied; in the back field, rye has been grown for 45 years in a row.

Crop rotation is the practice of growing a series of dissimilar or different types of crops in the same area in sequenced seasons. It is done so that the soil of farms is not used to only one type of nutrient. It helps in reducing soil erosion and increases soil fertility and crop yield.

Growing the same crop in the same place for many years in a row disproportionately depletes the soil of certain nutrients. With rotation, a crop that leaches the soil of one kind of nutrient is followed during the next growing season by a dissimilar crop that returns that nutrient to the soil or draws a different ratio of nutrients. In addition, crop rotation mitigates the buildup of pathogens and pests that often occurs when one species is continuously cropped, and can also improve soil structure and fertility by increasing biomass from varied root structures.

Crop rotation is used in both conventional and organic farming systems.

History

Agriculturalists have long recognized that suitable rotations – such as planting spring crops for livestock in place of grains for human consumption – make it possible to restore or to maintain a productive soil. Middle Eastern farmers practised crop rotation in 6000 BC without understanding the chemistry, alternately planting legumes and cereals. In the Bible, chapter 25 of the Book of Leviticus instructs the Israelites to observe a "Sabbath of the Land". Every seventh year they would not till, prune or even control insects. The Roman writer, Cato the Elder (234 – 149 BC), recommended that farmers "save carefully goat, sheep, cattle, and all other dung". From the times of Charlemagne (died 814), farmers in Europe transitioned from a two-field crop rotation to a three-field crop rotation. Under a two-field rotation, half the land was planted in a year, while the other half lay fallow. Then, in the next year, the two fields were reversed.

From the end of the Middle Ages until the 20th century, Europe's farmers practised three-year rotation, dividing available lands into three parts under three-field rotation. One section was planted in the autumn with rye or winter wheat, followed by spring oats or barley; the second section grew crops such as peas, lentils, or beans; and the third field was left fallow. The three fields were rotated in this manner so that every three years, a field would rest and be fallow. Under the two-field system, if one has a total of 600 acres (2.4 km²) of fertile land, one would only plant 300 acres. Under the new three-field rotation system, one would plant (and therefore harvest) 400 acres. But the additional crops had a more significant effect than mere quantitative productivity. Since the spring crops were mostly legumes, they increased the overall nutrition of the people of Northern Europe.

Farmers in the region of Waasland (in present-day northern Belgium) pioneered a four-field rotation in the early 16th century, and the British agriculturist Charles Townshend (1674-1738) popularised this system in the 18th century. The sequence of four crops (wheat, turnips, barley and clover), included a fodder crop and a grazing crop, allowing livestock to be bred year-round. The four-field crop rotation became a key development in the British Agricultural Revolution.

George Washington Carver (1860s - 1943) studied crop-rotation methods in the United States, teaching southern farmers to rotate soil-depleting crops like cotton with soil-enriching crops like peanuts and peas.

In the Green Revolution of the mod-20th century the traditional practice of crop rotation gave way in some parts of the world to the practice of supplementing the chemical inputs to the soil through topdressing with fertilizers, adding (for example) ammonium nitrate or urea and restoring soil pH with lime. Such practices aimed to increase yields, to prepare soil for specialist crops, and to reduce waste and inefficiency by simplifying planting and harvesting.

Crop Choice

A preliminary assessment of crop interrelationships can be found in how each crop: (1) contributes to soil organic matter (SOM) content, (2) provides for pest management, (3) manages deficient or excess nutrients, and (4) how it contributes to or controls for soil erosion.

Crop choice is often a related to the goal the farmer is looking to achieve with the rotation, which could be weed management, increasing available nitrogen in the soil, controlling for erosion, or increasing soil structure and biomass, to name a few. When discussing crop rotations, crops are classified in different ways depending on what quality is being assessed: by family, by nutrient needs/benefits, and/or by profitability (i.e. cash crop versus cover crop). For example, giving adequate attention to plant family is essential to mitigating pests and pathogens. However, many farmers have success managing rotations by planning sequencing and cover crops around desirable cash crops. The following is a simplified classification based on crop quality and purpose.

Row Crops

Many crops which are critical for the market, like vegetables, are row crops (that is, grown in tight rows). While often the most profitable for farmers, these crops are more taxing on the soil Row crops typically have low biomass and shallow roots: this means the plant contributes low residue to the surrounding soil and has limited effects on structure. With much of the soil around the plant is exposed to disruption by rainfall and traffic, fields with row crops experience faster break down of organic matter by microbes, leaving fewer nutrients for future plants.

In short, while these crops may be profitable for the farm, they are nutrient depleting. Crop rotation practices exist to strike a balance between short-term profitability and long-term productivity.

Legumes

A great advantage of crop rotation comes from the interrelationship of nitrogen fixing-crops with nitrogen demanding crops. Legumes, like alfalfa and clover, collect available nitrogen from the soil in nodules on their root structure. When the plant is harvested, the biomass of uncollected roots breaks down, making the stored nitrogen available to future crops. Legumes are also a valued green manure: a crop that collects nutrients and fixes them at soil depths accessible to future crops.

In addition, legumes have heavy tap roots that burrow deep into the ground, lifting soil for better tilth and absorption of water.

Grasses and Cereals

Cereal and grasses are frequent cover crops because of the many advantages they supply to soil quality and structure. The dense and far-reaching root systems give ample structure to surrounding soil and provide significant biomass for soil organic matter.

Grasses and cereals are key in weed management as they compete with undesired plants for soil space and nutrients.

Green Manure

Green manure is a crop that is mixed into the soil. Both nitrogen-fixing legumes and nutrient scavengers, like grasses, can be used as green manure. Green manure of legumes is an excellent source of nitrogen, especially for organic systems, however, legume biomass doesn't contribute to lasting soil organic matter like grasses do.

Planning a Rotation

There are numerous factors that must be taken into consideration when planning a crop rotation. Planning an effective rotation requires weighing fixed and fluctuating production circumstances, including, but not limited to: market, farm size, labor supply, climate, soil type, growing practices, etc. Moreover, a crop rotation must consider in what condition one crop will leave the soil for the succeeding crop and how one crop can be seeded with another crop. For example, a nitrogen-fixing crop, like a legume, should always proceed a nitrogen depleting one; similarly, a low residue crop (i.e. a crop with low biomass) should be offset with a high biomass cover crop, like a mixture of grasses and legumes.

There is no limit to the number of crops that can be used in a rotation, or the amount of time a rotation takes to complete. Decisions about rotations are made years prior, seasons prior, or even at the very last minute when an opportunity to increase profits or soil quality presents itself. In short, there is no singular formula for rotation, but many considerations to take into account.

Implementation

Crop rotation systems may be enriched by the influences of other practices such as the addition of livestock and manure, intercropping or multiple cropping, and organic management low in pesticides and synthetic fertilizers.

Incorporation of Livestock

Introducing livestock makes the most efficient use of critical sod and cover crops; livestock (through manure) are able to distribute the nutrients in these crops throughout the soil rather than removing nutrients from the farm through the sale of hay. In systems where use of farm livestock would violate reservations growers or consumers may have about animal exploitation, efforts are made to surrogate this input through livestock in the soil, namely worms and microorganisms.

In Sub-Saharan Africa, as animal husbandry becomes less of a nomadic practice many herders have begun integrating crop production into their practice. This is known as mixed farming, or the practice of crop cultivation with the incorporation of raising cattle, sheep and/or goats by the same economic entity, is increasingly common. This interaction between the animal, the land and the crops are being done on a small scale all across this region. Crop residues provide animal feed, while the animals provide manure for replenishing crop nutrients and draft power. Both processes are extremely important in this region of the world as it is expensive and logistically unfeasible to transport in synthetic fertilizers and large-scale machinery. As an additional benefit, the cattle, sheep and/or goat provide milk and can act as a cash crop in the times of economic hardship.

Organic Farming

Crop rotation is a required practice in order for a farm to receive organic certification in the United States. The "Crop Rotation Practice Standard" for the National Organic Program under the U.S. Code of Federal Regulations, section §205.205, states that:

Farmers are required to implement a crop rotation that maintains or builds soil organic matter, works to control pests, manages and conserves nutrients, and protects against erosion. Producers of perennial crops that aren't rotated may utilize other practices, such as cover crops, to maintain soil health.

In addition to lowering the need for inputs by controlling for pests and weeds and increasing available nutrients, crop rotation helps organic growers increase the amount of biodiversity on their farms. Biodiversity is also a requirement of organic certification, however, there are no rules in place to regulate or reinforce this standard. Increasing the biodiversity of crops has beneficial effects on the surrounding ecosystem and can host a greater diversity of fauna, insects, and beneficial microorganism in the soil.< Some studies point to increased nutrient availability from crop rotation under organic systems compared to conventional practices as organic practices are less likely to inhibit of beneficial microbes in soil organic matter.

While multiple cropping and intercropping benefit from many of the same principals as crop rotation, they do not satisfy the requirement under the NOP.

Intercropping

Multiple cropping systems, such as intercropping or companion planting, offer more diversity and complexity within the same season or rotation, for example the three sisters. An example of companion planting is the inter-planting of corn with pole beans and vining squash or pumpkins. In this system, the beans provide nitrogen; the corn provides support for the beans and a "screen" against squash vine borer; the vining squash provides a weed suppressive canopy and discourages corn-hungry raccoons.

Double-cropping is common where two crops, typically of different species, are grown sequentially in the same growing season, or where one crop (e.g. vegetable) is grown continuously with a cover crop (e.g. wheat). This is advantageous for small farms, who often cannot afford to leave cover crops to replenish the soil for extended periods of time, as larger farms can. When multiple cropping is implemented on small farms, these systems can maximize benefits of crop rotation on available land resources.

Benefits

Agronomists describe the benefits to yield in rotated crops as "The Rotation Effect". There are many found benefits of rotation systems: however, there is no specific scientific basis for the sometimes 10-25% yield increase in a crop grown in rotation versus monoculture. The factors related to the increase are simply described as alleviation of the negative factors of monoculture cropping systems. Explanations due to improved nutrition; pest, pathogen, and weed stress reduction; and improved soil structure have been found in some cases to be correlated, but causation has not been determined for the majority of cropping systems.

Other benefits of rotation cropping systems include production cost advantages. Overall financial risks are more widely distributed over more diverse production of crops and/or livestock. Less reliance is placed on purchased inputs and over time crops can maintain production goals with fewer inputs. This in tandem with greater short and long term yields makes rotation a powerful tool for improving agricultural systems.

Soil Organic Matter

The use of different species in rotation allows for increased soil organic matter (SOM), greater soil structure, and improvement of the chemical and biological soil environment for crops. With more SOM, water infiltration and retention improves, providing increased drought tolerance and decreased erosion.

Soil organic matter is a mix of decaying material from biomass with active microorganisms. Crop rotation, by nature, increases exposure to biomass from sod, green manure, and a various other plant debris. The reduced need for intensive tillage under crop rotation allows biomass aggregation to lead to greater nutrient retention and utilization, decreasing the need for added nutrients. With tillage, disruption and oxidation of soil creates a less conducive environment for diversity and proliferation of microorganisms in the soil. These microorganisms are what make nutrients available to plants. So, where "active" soil organic matter is a key to productive soil, soil with low microbial activity provides significantly fewer nutrients to plants; this is true even though the quantity of biomass left in the soil may be the same.

Soil microorganisms also decrease pathogen and pest activity through competition. In addition, plants produce root exudates and other chemicals which manipulate their soil environment as well as their weed environment. Thus rotation allows increased yields from nutrient availability but also alleviation of allelopathy and competitive weed environments.

Carbon Sequestration

Studies have shown that crop rotations greatly increase soil organic carbon (SOC) content, the main constituent of soil organic matter. Carbon, along with hydrogen and oxygen, is a macronutrient for plants. Highly diverse rotations spanning long periods of time have shown to be even more effective in increasing SOC, while soil disturbances (e.g. from tillage) are responsible for exponential decline in SOC levels. In Brazil, conservation to no-till methods combined with intensive crop rotations has been shown an SOC sequestration rate of 0.41 tonnes per hectare per year.

In addition to enhancing crop productivity, sequestration of atmospheric carbon has great implications in reducing rates of climate change by removing carbon dioxide from the air.

Nitrogen Fixing

Rotating crops adds nutrients to the soil. Legumes, plants of the family Fabaceae, for instance, have nodules on their roots which contain nitrogen-fixing bacteria called rhizobia. It therefore makes good sense agriculturally to alternate them with cereals (family Poaceae) and other plants that require nitrates.

Pathogen and Pest Control

Crop rotation is also used to control pests and diseases that can become established in the soil over time. The changing of crops in a sequence decreases the population level of pests by (1) interrupting pest life cycles and (2) interrupting pest habitat. Plants within the same taxonomic family tend to have similar pests and pathogens. By regularly changing crops and keeping the soil occupied by cover crops instead of lying fallow, pest cycles can be broken or limited, especially cycles that benefit from overwintering in residue. For example, root-knot nematode is a serious problem for some plants in warm climates and sandy soils, where it slowly builds up to high levels in the soil, and can severely damage plant productivity by cutting off circulation from the plant roots. Growing a crop that is not a host for root-knot nematode for one season greatly reduces the level of the nematode in the soil, thus making it possible to grow a susceptible crop the following season without needing soil fumigation.

This principle is of particular use in organic farming, where pest control must be achieved without synthetic pesticides.

Weed Management

Integrating certain crops, especially cover crops, into crop rotations is of particular value to weed management. These crops crowd out weed through competition. In addition, the sod and compost from cover crops and green manure slows the growth of what weeds are still able to make it through the soil, giving the crops further competitive advantage. By removing slowing the growth and proliferation of weeds while cover crops are cultivated, farmers greatly reduce the presence of weeds for future crops, including shallow rooted and row crops, which are less resistant to weeds. Cover crops are, therefore, considered conservation crops because they protect otherwise fallow land from becoming overrun with weeds.

This system has advantages over other common practices for weeds management, such as tillage. Tillage is meant to inhibit growth of weeds by overturning the soil; however, this has a countering effect of exposing weed seeds that may have gotten buried and burying valuable crop seeds. Under crop rotation, the number of viable seeds in the soil is reduced through the reduction of the weed population.

Preventing Soil Erosion

Crop rotation can significantly reduce the amount of soil lost from erosion by water. In areas that are highly susceptible to erosion, farm management practices such as zero and reduced tillage can be supplemented with specific crop rotation methods to reduce raindrop impact, sediment detachment, sediment transport, surface runoff, and soil loss.

Protection against soil loss is maximized with rotation methods that leave the greatest mass of crop stubble (plant residue left after harvest) on top of the soil. Stubble cover in contact with the soil minimizes erosion from water by reducing overland flow velocity, stream power, and thus the ability of the water to detach and transport sediment. Soil Erosion and Cill prevent the disruption and detachment of soil aggregates that cause macropores to block, infiltration to decline, and runoff to increase. This significantly improves the resilience of soils when subjected to periods of erosion and stress.

The effect of crop rotation on erosion control varies by climate. In regions under relatively consistent climate conditions, where annual rainfall and temperature levels are assumed, rigid crop rotations can produce sufficient plant growth and soil cover. In regions where climate conditions are less predictable, and unexpected periods of rain and drought may occur, a more flexible approach for soil cover by crop rotation is necessary. An opportunity cropping system promotes adequate soil cover under these erratic climate conditions. In an opportunity cropping system, crops are grown when soil water is adequate and there is a reliable sowing window. This form of cropping system is likely to produce better soil cover than a rigid crop rotation because crops are only sown under optimal conditions, whereas rigid systems are not necessarily sown in the best conditions available.

Crop rotations also affect the timing and length of when a field is subject to fallow. This is very important because depending on a particular region's climate, a field could be the most vulnerable to erosion when it is under fallow. Efficient fallow management is an essential part of reducing erosion in a crop rotation system. Zero tillage is a fundamental management practice that promotes crop stubble retention under longer unplanned fallows when crops cannot be planted. Such management practices that succeed in retaining suitable soil cover in areas under fallow will ultimately reduce soil loss.

Biodiversity

Increasing the biodiversity of crops has beneficial effects on the surrounding ecosystem and can host a greater diversity of fauna, insects, and beneficial microorganisms in the soil. Some studies point to increased nutrient availability from crop rotation under organic systems compared to conventional practices as organic practices are less likely to inhibit of beneficial microbes in soil organic matter, such as arbuscular mycorrhizae, which increase nutrient uptake in plants. Increasing biodiversity also increases the resilience of agro-ecological systems.

Farm Productivity

Crop rotation contributes to increased yields through improved soil nutrition. By requiring planting and harvesting of different crops at different times, more land can be farmed with the same amount of machinery and labour.

Risk Management

Different crops in the rotation can reduce the risks of adverse weather for the individual farmer.

Challenges

While crop rotation requires a great deal of planning, crop choice must respond to a number of fixed conditions (soil type, topography, climate, and irrigation) in addition to conditions that may change dramatically from year to the next (weather, market, labor supply). In this way, it is unwise to plan crops years in advance. Improper implementation of a crop rotation plan may lead to imbalances in the soil nutrient composition or a buildup of pathogens affecting a critical crop. The consequences of faulty rotation may take years to become apparent even to experienced soil scientists and can take just as long to correct.

Many challenges exist within the practices associated with crop rotation. For example, green manure from legumes can lead to an invasion of snails or slugs and the decay from green manure can occasionally suppress the growth of other crops.

Plant Breeding

The Yecoro wheat (right) cultivar is sensitive to salinity, plants resulting from a
hybrid cross with cultivar W4910 (left) show greater tolerance to high salinity

Plant breeding is the art and science of changing the traits of plants in order to produce desired characteristics. Plant breeding can be accomplished through many different techniques ranging from simply selecting plants with desirable characteristics for propagation, to more complex molecular techniques.

Plant breeding has been practiced for thousands of years, since near the beginning of human civilization. It is practiced worldwide by individuals such as gardeners and farmers, or by professional plant breeders employed by organizations such as government institutions, universities, crop-specific industry associations or research centers.

International development nation agencies believe that breeding new crops is important for ensuring food security by developing new varieties that are higher-yielding, disease resistant, drought-resistant or regionally adapted to different environments and growing conditions.

History

Plant breeding started with sedentary agriculture and particularly the domestication of the first agricultural plants, a practice which is estimated to date back 9,000 to 11,000 years. Initially early farmers simply selected food plants with particular desirable characteristics, and employed these as progenitors for subsequent generations, resulting in an accumulation of valuable traits over time.

Gregor Mendel's experiments with plant hybridization led to his establishing laws of inheritance. Once this work became well known, it formed the basis of the new science of genetics, which stimulated research by many plant scientists dedicated to improving crop production through plant breeding.

Modern plant breeding is applied genetics, but its scientific basis is broader, covering molecular biology, cytology, systematics, physiology, pathology, entomology, chemistry, and statistics (biometrics). It has also developed its own technology.

Classical Plant Breeding

One major technique of plant breeding is selection, the process of selectively propagating plants with desirable characteristics and eliminating or "culling" those with less desirable characteristics.

Another technique is the deliberate interbreeding (crossing) of closely or distantly related individuals to produce new crop varieties or lines with desirable properties. Plants are crossbred to introduce traits/genes from one variety or line into a new genetic background. For example, a mildew-resistant pea may be crossed with a high-yielding but susceptible pea, the goal of the cross being to introduce mildew resistance without losing the high-yield characteristics. Progeny from the cross would then be crossed with the high-yielding parent to ensure that the progeny were most like the high-yielding parent, (backcrossing). The progeny from that cross would then be tested for yield (selection, as described above) and mildew resistance and high-yielding resistant plants would be further developed. Plants may also be crossed with themselves to produce inbred varieties for breeding. Pollinators may be excluded through the use of pollination bags.

Classical breeding relies largely on homologous recombination between chromosomes to generate genetic diversity. The classical plant breeder may also make use of a number of *in vitro* techniques such as protoplast fusion, embryo rescue or mutagenesis to generate diversity and produce hybrid plants that would not exist in nature.

Traits that breeders have tried to incorporate into crop plants include:

1. Improved quality, such as increased nutrition, improved flavor, or greater beauty

2. Increased yield of the crop

3. Increased tolerance of environmental pressures (salinity, extreme temperature, drought)

4. Resistance to viruses, fungi and bacteria

5. Increased tolerance to insect pests

6. Increased tolerance of herbicides

7. Longer storage period for the harvested crop

Before World War II

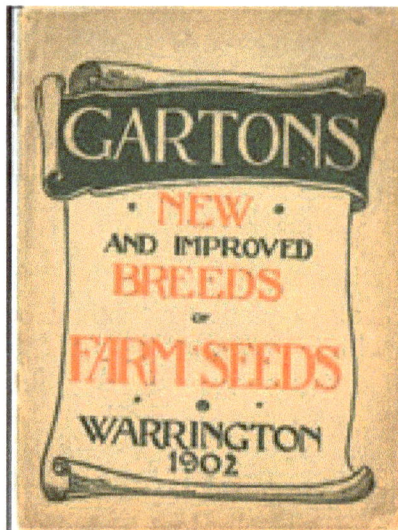

Garton's catalogue from 1902

Successful commercial plant breeding concerns were founded from the late 19th century. Gartons Agricultural Plant Breeders in England was established in the 1890s by John Garton, who was one of the first to commercialize new varieties of agricultural crops created through cross-pollination. The firm's first introduction was Abundance Oat, one of the first agricultural grain varieties bred from a *controlled* cross, introduced to commerce in 1892.

In the early 20th century, plant breeders realized that Mendel's findings on the non-random nature of inheritance could be applied to seedling populations produced through deliberate pollinations to predict the frequencies of different types. Wheat hybrids were bred to increase the crop production of Italy during the so-called "Battle for Grain" (1925–1940). Heterosis was explained by George Harrison Shull. It describes the tendency of the progeny of a specific cross to outperform both parents. The detection of the usefulness of heterosis for plant breeding has led to the development of inbred lines that reveal a heterotic yield advantage when they are crossed. Maize was the first species where heterosis was widely used to produce hybrids.

Statistical methods were also developed to analyze gene action and distinguish heritable variation from variation caused by environment. In 1933 another important breeding technique, cytoplasmic male sterility (CMS), developed in maize, was described by Marcus Morton Rhoades. CMS is a maternally inherited trait that makes the plant produce sterile pollen. This enables the production of hybrids without the need for labor-intensive detasseling.

These early breeding techniques resulted in large yield increase in the United States in the early 20th century. Similar yield increases were not produced elsewhere until after World War II, the Green Revolution increased crop production in the developing world in the 1960s.

After World War II

Following World War II a number of techniques were developed that allowed plant breeders to hybridize distantly related species, and artificially induce genetic diversity.

In vitro-culture of Vitis (grapevine), Geisenheim Grape Breeding Institute

When distantly related species are crossed, plant breeders make use of a number of plant tissue culture techniques to produce progeny from otherwise fruitless mating. Interspecific and intergeneric hybrids are produced from a cross of related species or genera that do not normally sexually reproduce with each other. These crosses are referred to as *Wide crosses*. For example, the cereal triticale is a wheat and rye hybrid. The cells in the plants derived from the first generation created from the cross contained an uneven number of chromosomes and as result was sterile. The cell division inhibitor colchicine was used to double the number of chromosomes in the cell and thus allow the production of a fertile line.

Failure to produce a hybrid may be due to pre- or post-fertilization incompatibility. If fertilization is possible between two species or genera, the hybrid embryo may abort before maturation. If this does occur the embryo resulting from an interspecific or intergeneric cross can sometimes be rescued and cultured to produce a whole plant. Such a method is referred to as Embryo Rescue. This technique has been used to produce new rice for Africa, an interspecific cross of Asian rice *(Oryza sativa)* and African rice *(Oryza glaberrima)*.

Hybrids may also be produced by a technique called protoplast fusion. In this case protoplasts are fused, usually in an electric field. Viable recombinants can be regenerated in culture.

Chemical mutagens like EMS and DMS, radiation and transposons are used to generate mutants with desirable traits to be bred with other cultivars - a process known as *Mutation Breeding*. Classical plant breeders also generate genetic diversity within a species by exploiting a process called somaclonal variation, which occurs in plants produced from tissue culture, particularly plants derived from callus. Induced polyploidy, and the addition or removal of chromosomes using a technique called chromosome engineering may also be used.

When a desirable trait has been bred into a species, a number of crosses to the favored parent are made to make the new plant as similar to the favored parent as possible. Returning to the example of the mildew resistant pea being crossed with a high-yielding but susceptible pea, to make the mildew resistant progeny of the cross most like the high-yielding parent, the progeny will be crossed back to that parent for several generations. This process removes most of the genetic contribution of the mildew resistant parent. Classical breeding is therefore a cyclical process.

With classical breeding techniques, the breeder does not know exactly what genes have been introduced to the new cultivars. Some scientists therefore argue that plants produced by classical breeding methods should undergo the same safety testing regime as genetically modified plants. There have been instances where plants bred using classical techniques have been unsuitable for human consumption, for example the poison solanine was unintentionally increased to unacceptable levels in certain varieties of potato through plant breeding. New potato varieties are often screened for solanine levels before reaching the marketplace.

Modern Plant Breeding

Modern plant breeding may use techniques of molecular biology to select, or in the case of genetic modification, to insert, desirable traits into plants. Application of biotechnology or molecular biology is also known as molecular breeding.

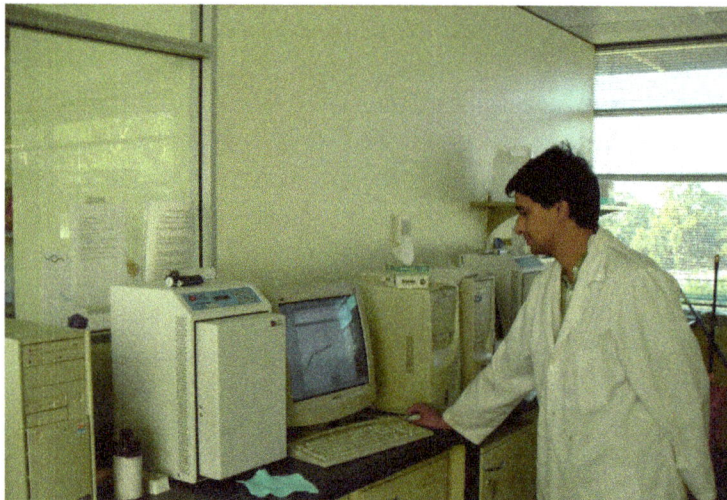

Modern facilities in molecular biology have converted classical plant breeding to molecular plant breeding

Steps of Plant Breeding

The following are the major activities of plant breeding:

1. Collection of variation

2. Selection

3. Evaluation

4. Release

5. Multiplication

6. Distribution of the new variety

Marker Assisted Selection

Sometimes many different genes can influence a desirable trait in plant breeding. The use of tools such as molecular markers or DNA fingerprinting can map thousands of genes. This allows plant breeders to screen large populations of plants for those that possess the trait of interest. The screening is based on the presence or absence of a certain gene as determined by laboratory procedures, rather than on the visual identification of the expressed trait in the plant.

Reverse Breeding and Doubled Haploids (DH)

A method for efficiently producing homozygous plants from a heterozygous starting plant, which has all desirable traits. This starting plant is induced to produce doubled haploid from haploid cells, and later on creating homozygous/doubled haploid plants from those cells. While in natural offspring genetic recombination occurs and traits can be unlinked from each other, in doubled haploid cells and in the resulting DH plants recombination is no longer an issue. There, a recombination between two corresponding chromosomes does not lead to un-linkage of alleles or traits, since it just leads to recombination with its identical copy. Thus, traits on one chromosome stay linked. Selecting those offspring having the desired set of chromosomes and crossing them will result in a final F1 hybrid plant, having exactly the same set of chromosomes, genes and traits as the starting hybrid plant. The homozygous parental lines can reconstitute the original heterozygous plant by crossing, if desired even in a large quantity. An individual heterozygous plant can be converted into a heterozygous variety (F1 hybrid) without the necessity of vegetative propagation but as the result of the cross of two homozygous/doubled haploid lines derived from the originally selected plant. patent

Genetic Modification

Genetic modification of plants is achieved by adding a specific gene or genes to a plant, or by knocking down a gene with RNAi, to produce a desirable phenotype. The plants resulting from adding a gene are often referred to as transgenic plants. If for genetic modification genes of the species or of a crossable plant are used under control of their native promoter, then they are called cisgenic plants. Sometimes genetic modification can produce a plant with the desired trait or traits faster than classical breeding because the majority of the plant's genome is not altered.

To genetically modify a plant, a genetic construct must be designed so that the gene to be added or removed will be expressed by the plant. To do this, a promoter to drive transcription and a termination sequence to stop transcription of the new gene, and the gene or genes of interest must be introduced to the plant. A marker for the selection of transformed plants is also included. In the laboratory, antibiotic resistance is a commonly used marker: Plants that have been successfully transformed will grow on media containing antibiotics; plants that have not been transformed will die. In some instances markers for selection are removed by backcrossing with the parent plant prior to commercial release.

The construct can be inserted in the plant genome by genetic recombination using the bacteria *Agrobacterium tumefaciens* or *A. rhizogenes*, or by direct methods like the gene gun or micro-injection. Using plant viruses to insert genetic constructs into plants is also a possibility, but the technique is limited by the host range of the virus. For example, Cauliflower mosaic virus (CaMV) only infects cauliflower and related species. Another limitation of viral vectors is that the virus is not usually passed on the progeny, so every plant has to be inoculated.

The majority of commercially released transgenic plants are currently limited to plants that have introduced resistance to insect pests and herbicides. Insect resistance is achieved through incorporation of a gene from *Bacillus thuringiensis* (Bt) that encodes a protein that is toxic to some insects. For example, the cotton bollworm, a common cotton pest, feeds on Bt cotton it will ingest the toxin and die. Herbicides usually work by binding to certain plant enzymes and inhibiting their action. The enzymes that the herbicide inhibits are known as the herbicides *target site*. Herbicide resistance can be engineered into crops by expressing a version of *target site* protein that is not inhibited by the herbicide. This is the method used to produce glyphosate resistant crop plants.

Genetic modification of plants that can produce pharmaceuticals (and industrial chemicals), sometimes called *pharming*, is a rather radical new area of plant breeding.

Issues and Concerns

Modern plant breeding, whether classical or through genetic engineering, comes with issues of concern, particularly with regard to food crops. The question of whether breeding can have a negative effect on nutritional value is central in this respect. Although relatively little direct research in this area has been done, there are scientific indications that, by favoring certain aspects of a plant's development, other aspects may be retarded. A study published in the *Journal of the American College of Nutrition* in 2004, entitled *Changes in USDA Food Composition Data for 43 Garden Crops, 1950 to 1999*, compared nutritional analysis of vegetables done in 1950 and in 1999, and found substantial decreases in six of 13 nutrients measured, including 6% of protein and 38% of riboflavin. Reductions in calcium, phosphorus, iron and ascorbic acid were also found. The study, conducted at the Biochemical Institute, University of Texas at Austin, concluded in summary: *"We suggest that any real declines are generally most easily explained by changes in cultivated varieties between 1950 and 1999, in which there may be trade-offs between yield and nutrient content."*

The debate surrounding genetically modified food during the 1990s peaked in 1999 in terms of media coverage and risk perception, and continues today - for example, *"Germany has thrown its weight behind a growing European mutiny over genetically modified crops by banning the planting of a widely grown pest-resistant corn variety."* The debate encompasses the ecological impact of genetically modified plants, the safety of genetically modified food and concepts used for safety evaluation like substantial equivalence. Such concerns are not new to plant breeding. Most countries have regulatory processes in place to help ensure that new crop varieties entering the marketplace are both safe and meet farmers' needs. Examples include variety registration, seed schemes, regulatory authorizations for GM plants, etc.

Plant breeders' rights is also a major and controversial issue. Today, production of new varieties is dominated by commercial plant breeders, who seek to protect their work and collect royalties through national and international agreements based in intellectual property rights. The range of related issues is complex. In the simplest terms, critics of the increasingly restrictive regulations argue that, through a combination of technical and economic pressures, commercial breeders are reducing biodiversity and significantly constraining individuals (such as farmers) from developing and trading seed on a regional level. Efforts to strengthen breeders' rights, for example, by lengthening periods of variety protection, are ongoing.

When new plant breeds or cultivars are bred, they must be maintained and propagated. Some plants are propagated by asexual means while others are propagated by seeds. Seed propagated cultivars require specific control over seed source and production procedures to maintain the integrity of the plant breeds results. Isolation is necessary to prevent cross contamination with related plants or the mixing of seeds after harvesting. Isolation is normally accomplished by planting distance but in certain crops, plants are enclosed in greenhouses or cages (most commonly used when producing F1 hybrids.)

Role of Plant Breeding in Organic Agriculture

Critics of organic agriculture claim it is too low-yielding to be a viable alternative to conventional agriculture. However, part of that poor performance may be the result of growing poorly adapted varieties. It is estimated that over 95% of organic agriculture is based on conventionally adapted varieties, even though the production environments found in organic vs. conventional farming systems are vastly different due to their distinctive management practices. Most notably, organic farmers have fewer inputs available than conventional growers to control their production environments. Breeding varieties specifically adapted to the unique conditions of organic agriculture is critical for this sector to realize its full potential. This requires selection for traits such as:

- Water use efficiency
- Nutrient use efficiency (particularly nitrogen and phosphorus)
- Weed competitiveness
- Tolerance of mechanical weed control
- Pest/disease resistance
- Early maturity (as a mechanism for avoidance of particular stresses)
- Abiotic stress tolerance (i.e. drought, salinity, etc...)

Currently, few breeding programs are directed at organic agriculture and until recently those that did address this sector have generally relied on indirect selection (i.e. selection in conventional environments for traits considered important for organic agriculture). However, because the difference between organic and conventional environments is large, a given genotype may perform very differently in each environment due to an interaction between genes and the environment. If this interaction is severe enough, an important trait required for the organic environment may not be revealed in the conventional environment, which can result in the selection of poorly adapted individuals. To ensure the most adapted varieties are identified, advocates of organic breeding now promote the use of direct selection (i.e. selection in the target environment) for many agronomic traits.

There are many classical and modern breeding techniques that can be utilized for crop improvement in organic agriculture despite the ban on genetically modified organisms. For instance, controlled crosses between individuals allow desirable genetic variation to be recombined and transferred to seed progeny via natural processes. Marker assisted selection can also be employed as a diagnostics tool to facilitate selection of progeny who possess the desired trait(s), greatly speeding up the breeding process. This technique has proven particularly useful for the introgression of resistance genes into new backgrounds, as well as the efficient selection of many resistance genes pyramided into a single individual. Unfortunately, molecular markers are not currently available for many important traits, especially complex ones controlled by many genes.

Addressing Global Food Security Through Plant Breeding

For future agriculture to thrive there are necessary changes which must be made in accordance to arising global issues. These issues are arable land, harsh cropping conditions and food security which involves, being able to provide the world population with food containing sufficient nutrients. These crops need to be able to mature in several environments allowing for worldwide access, this is involves issues such as drought tolerance. These global issues are achievable through the process of plant breeding, as it offers the ability to select specific genes allowing the crop to perform at a level which yields the desired results.

Minimal Land Degradation

Land degradation is a major issue, as it can negatively impact the capability of the land to be productive. Poor agricultural management has a huge impact on the degradation of soil worldwide and it is Africa and Asia that are most affected. Through education and development of modified plants, these statistics can be reduced and agricultural land can become more productive. Plant breeding allows for an increase in yield with out the extra strain on the land. The genetically modified, Bt white maize, was introduced to South Africa and was surveyed in 33 large commercial farms and 368 small landholders properties and in both cases a higher yield was recorded.

Increased Yield Without Expansion

With an increasing population, the production of food needs to increase with it. It is estimated that a 70% increase in food production is needed by 2050 in order to meet the Declaration of the World Summit on Food Security. But with the natural degradation of agricultural land, simply planting more crops is no longer a viable option. Therefore, new varieties of plants need to be developed through plant breeding that generates an increase of yield without relying on an increase in land area. An example of this can be seen in Asia, where food production per capita has increased twofold. This has been achieved through not only the use of fertilisers, but through the use of better crops that have been specifically designed for the area.

Breeding for Increased Nutritional Value

Plant breeding can contribute to global food security as it is a cost-effective tool for increasing nutritional value of forage and crops. Improvements in nutritional value for forage crops from the use of analytical chemistry and rumen fermentation technology have been recorded since 1960;

this science and technology gave breeders the ability to screen thousands of samples within a small amount of time, meaning breeders could identify a high performing hybrid quicker. The main area genetic increases were made was in vitro dry matter digestibility (IVDMD) resulting in 0.7-2.5% increase, at just 1% increase in IVDMD a single Bos Taurus also known as beef cattle reported 3.2% increase in daily gains. This improvement indicates plant breeding is an essential tool in gearing future agriculture to perform at a more advanced level.

Breeding for Tolerance

Plant breeding of hybrid crops has become extremely popular worldwide in an effort to combat the harsh environment. With long periods of drought and lack of water or nitrogen stress tolerance has become a significant part of agriculture. Plant breeders have focused on identifying crops which will ensure crops perform under these conditions; a way to achieve this is finding strains of the crop that is resistance to drought conditions with low nitrogen. It is evident from this that plant breeding is vital for future agriculture to survive as it enables farmers to produce stress resistant crops hence improving food security.

Participatory Plant Breeding

The development of agricultural science, with phenomenon like the Green Revolution arising, have left millions of farmers in developing countries, most of whom operate small farms under unstable and difficult growing conditions, in a precarious situation. The adoption of new plant varieties by this group has been hampered by the constraints of poverty and the international policies promoting an industrialized model of agriculture. Their response has been the creation of a novel and promising set of research methods collectively known as participatory plant breeding. Participatory means that farmers are more involved in the breeding process and breeding goals are defined by farmers instead of international seed companies with their large-scale breeding programs. Farmers' groups and NGOs, for example, may wish to affirm local people's rights over genetic resources, produce seeds themselves, build farmers' technical expertise, or develop new products for niche markets, like organically grown food.

List of Notable Plant Breeders

- Gartons Agricultural Plant Breeders
- Gregor Mendel
- Keith Downey
- Luther Burbank
- Nazareno Strampelli
- Niels Ebbesen Hansen
- Edger McFadden
- Norman Borlaug

Shifting Cultivation

Slash-and-burn based shifting cultivation is a widespread historical practice in southeast Asia.
Above is a satellite image of Sumatra and Borneo showing shift cultivation fires from October 2006.

Shifting cultivation is an agricultural system in which plots of land are cultivated temporarily, then abandoned and allowed to revert to their natural vegetation while the cultivator moves on to another plot. The period of cultivation is usually terminated when the soil shows signs of exhaustion or, more commonly, when the field is overrun by weeds. The length of time that a field is cultivated is usually shorter than the period over which the land is allowed to regenerate by lying fallow. This technique is often used in LEDCs (Less Economically Developed Countries) or LICs (Low Income Countries).

Of these cultivators, many use a practice of slash-and-burn as one element of their farming cycle. Others employ land clearing without any burning, and some cultivators are purely migratory and do not use any cyclical method on a given plot. Sometimes no slashing at all is needed where regrowth is purely of grasses, an outcome not uncommon when soils are near exhaustion and need to lie fallow. In shifting agriculture, after two or three years of producing vegetable and grain crops on cleared land, the migrants abandon it for another plot. Trees and bushes are cleared by slashing, and the remaining vegetation is burnt. The ashes add potash to the soil. Then the seeds are sown after the rains

Advantages of Slash-and-burn Method

Slash-and-burn is a very sustainable technique. It differs a lot from commercial farming, because once the trees are burned, there is very fertile fine ash that deposits along the humus, meaning that by the time the other fields are burned, the soil has time to reassemble nutriments, in order to make cultural activity possible. Although slash-and-burn is a very useful technique, there are other ways of fertilizing soil. By planting beans, the soil will regenerate much faster, due to the production of nitrogen in their roots.

Political Ecology of Shifting Cultivation

Shifting cultivation is a form of agriculture or a cultivation system, in which, at any particular point in time, a minority of 'fields' are in cultivation and a majority are in various stages of natural re-growth. Over time, fields are cultivated for a relatively short time, and allowed to recover, or are fallowed, for a relatively long time. Eventually a previously cultivated field will be cleared of the natural vegetation and planted in crops again. Fields in established and stable shifting cultivation systems are cultivated and fallowed cyclically.This type of farming is called jhumming in India.

Fallow fields are not unproductive. During the fallow period, shifting cultivators use the successive vegetation species widely for timber for fencing and construction, firewood, thatching, ropes, clothing, tools, carrying devices and medicines. It is common for fruit and nut trees to be planted in fallow fields to the extent that parts of some fallows are in fact orchards. Soil-enhancing shrub or tree species may be planted or protected from slashing or burning in fallows. Many of these species have been shown to fix nitrogen. Fallows commonly contain plants that attract birds and animals and are important for hunting. But perhaps most importantly, tree fallows protect soil against physical erosion and draw nutrients to the surface from deep in the soil profile.

The relationship between the time the land is cultivated and the time it is fallowed are critical to the stability of shifting cultivation systems. These parameters determine whether or not the shifting cultivation system as a whole suffers a net loss of nutrients over time. A system in which there is a net loss of nutrients with each cycle will eventually lead to a degradation of resources unless actions are taken to arrest the losses. In some cases soil can be irreversibly exhausted (including erosion as well as nutrient loss) in less than a decade.

The longer a field is cropped, the greater the loss of soil organic matter, cation-exchange-capacity and in nitrogen and phosphorus, the greater the increase in acidity, the more likely soil porosity and infiltration capacity is reduced and the greater the loss of seeds of naturally occurring plant species from soil seed banks. In a stable shifting cultivation system, the fallow is long enough for the natural vegetation to recover to the state that it was in before it was cleared, and for the soil to recover to the condition it was in before cropping began. During fallow periods soil temperatures are lower, wind and water erosion is much reduced, nutrient cycling becomes closed again, nutrients are extracted from the subsoil, soil fauna decreases, acidity is reduced, soil structure, texture and moisture characteristics improve and seed banks are replenished.

The secondary forests created by shifting cultivation are commonly richer in plant and animal resources useful to humans than primary forests, even though they are much less bio-diverse. Shifting cultivators view the forest as an agricultural landscape of fields at various stages in a regular cycle. People unused to living in forests cannot see the fields for the trees. Rather they perceive an apparently chaotic landscape in which trees are cut and burned randomly and so they characterise shifting cultivation as ephemeral or 'pre-agricultural', as 'primitive' and as a stage to be progressed beyond. Shifting agriculture is none of these things. Stable shifting cultivation systems are highly variable, closely adapted to micro-environments and are carefully managed by farmers during both the cropping and fallow stages. Shifting cultivators may possess a highly developed knowledge and understanding of their local environments and of the crops and native plant species they exploit. Complex and highly adaptive land tenure systems sometimes exist under shifting cultivation. Introduced crops for food and as cash have been skillfully integrated into some shifting cultivation systems.

Shifting Cultivation in Europe

Shifting cultivation was still being practised as a viable and stable form of agriculture in many parts of Europe and east into Siberia at the end of the 19th century and in some places well into the 20th century. In the Ruhr in the late 1860s a forest-field rotation system known as *Reutberg-wirtschaft* was using a 16-year cycle of clearing, cropping and fallowing with trees to produce bark for tanneries, wood for charcoal and rye for flour (Darby 1956, 200). Swidden farming was practised in Siberia at least until the 1930s, using specially selected varieties of "swidden-rye" (Steensberg 1993, 98). In Eastern Europe and Northern Russia the main swidden crops were turnips, barley, flax, rye, wheat, oats, radishes and millet. Cropping periods were usually one year, but were extended to two or three years on very favourable soils. Fallow periods were between 20 and 40 years (Linnard 1970, 195). In Finland in 1949, Steensberg (1993, 111) observed the clearing and burning of a 60,000 square metres (15 acres) swidden 440 km north of Helsinki. Birch and pine trees had been cleared over a period of a year and the logs sold for cash. A fallow of alder (Alnus) was encouraged to improve soil conditions. After the burn, turnip was sown for sale and for cattle feed. Shifting cultivation was disappearing in this part of Finland because of a loss of agricultural labour to the industries of the towns. Steensberg (1993, 110-152) provides eye-witness descriptions of shifting cultivation being practised in Sweden in the 20th century, and in Estonia, Poland, the Caucasus, Serbia, Bosnia, Hungary, Switzerland, Austria and Germany in the 1930s to the 1950s.

That these agricultural practices survived from the Neolithic into the middle of the 20th century amidst the sweeping changes that occurred in Europe over that period, suggests they were adaptive and in themselves, were not massively destructive of the environments in which they were practised. This raises the question: if shifting cultivation did not lead to the disappearance of European forests, what did?

The earliest written accounts of forest destruction in Southern Europe begin around 1000 BC in the histories of Homer, Thucydides and Plato and in Strabo's Geography. Forests were exploited for ship building, and urban development, the manufacture of casks, pitch and charcoal, as well as being cleared for agriculture. The intensification of trade and as a result of warfare, increased the demand for ships which were manufactured completely from forest products. Although goat herding is singled out as an important cause of environmental degradation, a more important cause of forest destruction was the practice in some places of granting ownership rights to those who clear felled forests and brought the land into permanent cultivation. Evidence that circumstances other than agriculture were the major causes for forest destruction was the recovery of tree cover in many parts of the Roman empire from 400 BC to around 500 AD following the collapse of Roman economy and industry. Darby observes that by 400 AD "land that had once been tilled became derelict and overgrown" and quotes Lactantius who wrote that in many places "cultivated land became forest" (Darby 1956, 186). The other major cause of forest destruction in the Mediterranean environment with its hot dry summers were wild fires that became more common following human interference in the forests.

In Central and Northern Europe the use of stone tools and fire in agriculture is well established in the palynological and archaeological record from the Neolithic. Here, just as in Southern Europe, the demands of more intensive agriculture and the invention of the plough, trading, mining and smelting, tanning, building and construction in the growing towns and constant warfare, including the demands of naval shipbuilding, were more important forces behind the destruction of the forests than was shifting cultivation.

By the Middle Ages in Europe, large areas of forest were being cleared and converted into arable land in association with the development of feudal tenurial practices. From the 16th to the 18th centuries, the demands of iron smelters for charcoal, increasing industrial developments and the discovery and expansion of colonial empires as well as incessant warfare that increased the demand for shipping to levels never previously reached, all combined to deforest Europe. With the loss of the forest, so shifting cultivation became restricted to the peripheral places of Europe, where permanent agriculture was uneconomic, transport costs constrained logging or terrain prevented the use of draught animals or tractors. It has disappeared from even these refuges since 1945, as agriculture has become increasingly capital intensive, rural areas have become depopulated and the remnant European forests themselves have been revalued economically and socially.

Simple Societies, Shifting Cultivation and Environmental Change

Shifting cultivation in Indonesia. A new crop is sprouting through the burnt soil.

A growing body of palynological evidence finds that simple human societies brought about extensive changes to their environments before the establishment of any sort of state, feudal or capitalist, and before the development of large scale mining, smelting or shipbuilding industries. In these societies agriculture was the driving force in the economy and shifting cultivation was the most common type of agriculture practiced. By examining the relationships between social and economic change and agricultural change in these societies, insights can be gained on contemporary social and economic change and global environment change, and the place of shifting cultivation in those relationship.

As early as 1930 questions about relationships between the rise and fall of the Mayan civilization of the Yucatán Peninsula and shifting cultivation were raised and continue to be debated today. Archaeological evidence suggests the development of Mayan society and economy began around 250 AD. A mere 700 years later it reached its apogee, by which time the population may have reached 2,000,000 people. There followed a precipitous decline that left the great cities and ceremonial centres vacant and overgrown with jungle vegetation. The causes of this decline are uncertain; but warfare and the exhaustion of agricultural land are commonly cited (Meggers 1954; Dumond 1961; Turner 1974). More recent work suggests the Maya may have, in suitable places, developed irrigation systems and more intensive agricultural practices (Humphries 1993).

Similar paths appear to have been followed by Polynesian settlers in New Zealand and the Pacific Islands, who within 500 years of their arrival around 1100 AD turned substantial areas from forest into scrub and fern and in the process caused the elimination of numerous species of birds and animals (Kirch and Hunt 1997). In the restricted environments of the Pacific islands, including Fiji and Hawaii, early extensive erosion and change of vegetation is presumed to have been caused by shifting cultivation on slopes. Soils washed from slopes were deposited in valley bottoms as a rich, swampy alluvium. These new environments were then exploited to develop intensive, irrigated fields. The change from shifting cultivation to intensive irrigated fields occurred in association with a rapid growth in population and the development of elaborate and highly stratified chiefdoms (Kirch 1984). In the larger, temperate latitude, islands of New Zealand the presumed course of events took a different path. There the stimulus for population growth was the hunting of large birds to extinction, during which time forests in drier areas were destroyed by burning, followed the development of intensive agriculture in favorable environments, based mainly on sweet potato (Ipomoea batatas) and a reliance on the gathering of two main wild plant species in less favorable environments. These changes, as in the smaller islands, were accompanied by population growth, the competition for the occupation of the best environments, complexity in social organization, and endemic warfare (Anderson 1997).

The record of humanly induced changes in environments is longer in New Guinea than in most places. Agricultural activities probably began 5,000 to 9,000 years ago. However, the most spectacular changes, in both societies and environments, are believed to have occurred in the central highlands of the island within the last 1,000 years, in association with the introduction of a crop new to New Guinea, the sweet potato (Golson 1982a; 1982b). One of the most striking signals of the relatively recent intensification of agriculture is the sudden increase in sedimentation rates in small lakes.

The root question posed by these and the numerous other examples that could be cited of simple societies that have intensified their agricultural systems in association with increases in population and social complexity is not whether or how shifting cultivation was responsible for the extensive changes to landscapes and environments. Rather it is why simple societies of shifting cultivators in the tropical forest of Yucatán, or the highlands of New Guinea, began to grow in numbers and to develop stratified and sometimes complex social hierarchies?

At first sight, the greatest stimulus to the intensification of a shifting cultivation system is a growth in population. If no other changes occur within the system, for each extra person to be fed from the system, a small extra amount of land must be cultivated. The total amount of land available is the land being presently cropped and all of the land in fallow. If the area occupied by the system is not expanded into previously unused land, then either the cropping period must be extended or the fallow period shortened.

At least two problems exist with the population growth hypothesis. First, population growth in most pre-industrial shifting cultivator societies has been shown to be very low over the long term. Second, no human societies are known where people work only to eat. People engage in social relations with each other and agricultural produce is used in the conduct of these relationships.

These relationships are the focus of two attempts to understand the nexus between human societies and their environments, one an explanation of a particular situation and the other a general exploration of the problem.

Feedback Loops

In a study of the Duna in the Southern Highlands of New Guinea, a group in the process of moving from shifting cultivation into permanent field agriculture post sweet potato, Modjeska (1982) argued for the development of two "self amplifying feed back loops" of ecological and social causation. The trigger to the changes was very slow population growth and the slow expansion of agriculture to meet the demands of this growth. This set in motion the first feedback loop, the "use-value" loop. As more forest was cleared there was a decline in wild food resources and protein produced from hunting, which was substituted for by an increase in domestic pig raising. An increase in domestic pigs required a further expansion in agriculture. The greater protein available from the larger number of pigs increased human fertility and survival rates and resulted in faster population growth.

The outcome of the operation of the two loops, one bringing about ecological change and the other social and economic change, is an expanding and intensifying agricultural system, the conversion of forest to grassland, a population growing at an increasing rate and expanding geographically and a society that is increasing in complexity and stratification.

Resources are Cultural Appraisals

The second attempt to explain the relationships between simple agricultural societies and their environments is that of Ellen (1982, 252-270). Ellen does not attempt to separate use-values from social production. He argues that almost all of the materials required by humans to live (with perhaps the exception of air) are obtained through social relations of production and that these relations proliferate and are modified in numerous ways. The values that humans attribute to items produced from the environment arise out of cultural arrangements and not from the objects themselves, a restatement of Carl Sauer's dictum that "resources are cultural appraisals". Humans frequently translate actual objects into culturally conceived forms, an example being the translation by the Duna of the pig into an item of compensation and redemption. As a result, two fundamental processes underlie the ecology of human social systems: First, the obtaining of materials from the environment and their alteration and circulation through social relations, and second, giving the material a value which will affect how important it is to obtain it, circulate it or alter it. Environmental pressures are thus mediated through social relations.

Transitions in ecological systems and in social systems do not proceed at the same rate. The rate of phylogenetic change is determined mainly by natural selection and partly by human interference and adaptation, such as for example, the domestication of a wild species. Humans however have the ability to learn and to communicate their knowledge to each other and across generations. If most social systems have the tendency to increase in complexity they will, sooner or later, come into conflict with, or into "contradiction" (Friedman 1979, 1982) with their environments. What happens around the point of "contradiction" will determine the extent of the environmental degradation that will occur. Of particular importance is the ability of the society to change, to invent or to innovate technologically and sociologically, in order to overcome the "contradiction" without incurring continuing environmental degradation, or social disintegration.

An economic study of what occurs at the points of conflict with specific reference to shifting cultivation is that of Esther Boserup (1965). Boserup argues that low intensity farming, extensive shifting cultivation for example, has lower labor costs than more intensive farming systems. This assertion remains controversial. She also argues that given a choice, a human group will always choose the technique which has the lowest absolute labor cost rather than the highest yield. But at the point of conflict, yields will have become unsatisfactory. Boserup argues, contra Malthus, that rather than population always overwhelming resources, that humans will invent a new agricultural technique or adopt an existing innovation that will boost yields and that is adapted to the new environmental conditions created by the degradation which has occurred already, even though they will pay for the increases in higher labor costs. Examples of such changes are the adoption of new higher yielding crops, the exchanging of a digging stick for a hoe, or a hoe for a plough, or the development of irrigation systems. The controversy over Boserup's proposal is in part over whether intensive systems are more costly in labor terms, and whether humans will bring about change in their agricultural systems before environmental degradation forces them to.

Shifting Cultivation in the Contemporary World and Global Environmental Change

Contemporary Shifting Cultivation Practice

Sumatra, Indonesia

Rio Xingu, Brazil

Santa Cruz, Bolivia

Kasempa, Zambia

The estimated rate of deforestation in Southeast Asia in 1990 was 34,000 km² per year (FAO 1990, quoted in Potter 1993). In Indonesia alone it was estimated 13,100 km² per year were being lost, 3,680 km² per year from Sumatra and 3,770 km² from Kalimantan, of which 1,440 km² were due to the fires of 1982 to 1983. Since those estimates were made huge fires have ravaged Indonesian forests during the 1997 to 1998 El Niño associated drought.

Interdisciplinary Project

Shifting cultivation used to be the backbone of smallholder agriculture throughout the tropics, but today it is abandoned in many places in favor of large scale cash crop production – e.g. for biofuels. The extent of these changes is not well documented because shifting cultivation land rarely appears on official maps and census data seldom identifies shifting cultivators. Moreover, the consequences of these changes for livelihoods (e.g. food security) are not well known. The aim of this project is to analyze the extent and consequences of change in shifting cultivation by combining meta-analyses of existing studies and census data with case studies in selected areas. This interdisciplinary project focuses on:

1. Trends in change in shifting cultivation landscapes and demography; and

2. Changes in livelihoods due to these changes.

The project will compile data for eight countries (Mexico, Brazil, Laos, Vietnam, Malaysia, Thailand, Zambia and Tanzania) and the outcome is expected to be relevant to planning and policy-making on land and forest management.

Shifting cultivation was assessed by the FAO to be one a causes of deforestation while logging was not. The apparent discrimination against shifting cultivators caused a confrontation between FAO and environmental groups, who saw the FAO supporting commercial logging interests against the rights of indigenous people (Potter 1993, 108). Other independent studies of the problem note that despite lack of government control over forests and the dominance of a political elite in the logging industry, the causes of deforestation are more complex. The loggers have provided paid employment to former subsistence farmers. One of the outcomes of cash incomes has been rapid population growth among indigenous groups of former shifting cultivators that has placed pressure on their traditional long fallow farming systems. Many farmers have taken advantage of the improved road access to urban areas by planting cash crops, such as rubber or pepper as noted above. Increased cash incomes often are spent on chain saws, which have enabled larger areas to be cleared for cultivation. Fallow periods have been reduced and cropping periods extended. Serious poverty elsewhere in the country has brought thousands of land hungry settlers into the cut over forests along the logging roads. The settlers practice what appears to be shifting cultivation but which is in fact a one-cycle slash and burn followed by continuous cropping, with no intention to long fallow. Clearing of trees and the permanent cultivation of fragile soils in a tropical environment with little attempt to replace lost nutrients may cause rapid degradation of the fragile soils.

The loss of forest in Indonesia, Thailand, and the Philippines during the 1990s was preceded by major ecosystem disruptions in Vietnam, Laos and Cambodia in the 1970s and 1980s caused by warfare. Forests were sprayed with defoliants, thousands of rural forest dwelling people uproots from their homes and moved and roads driven into previously isolated areas. The loss of the tropical forests of Southeast Asia is the particular outcome of the general possible outcomes described by Ellen when small local ecological and social systems become part of larger system. When the previous relatively stable ecological relationships are destabilized, degradation can occur rapidly. Similar descriptions of the loss of forest and destruction of fragile ecosystems could be provided from the Amazon Basin, by large scale state sponsored colonization forest land (Becker 1995, 61) or from the Central Africa where what endemic armed conflict is destabilizing rural settlement and farming communities on a massive scale.

Comparison with Other Ecological Phenomena

In the tropical developing world, shifting cultivation in its many diverse forms, remains a pervasive practice. Shifting cultivation was one of the very first forms of agriculture practiced by humans and its survival into the modern world suggests that it is a flexible and highly adaptive means of production. However, it is also a grossly misunderstood practice. Many casual observers cannot see past the clearing and burning of standing forest and do not perceive often ecologically stable cycles of cropping and fallowing. Nevertheless, shifting cultivation systems are particularly susceptible to rapid increases in population and to economic and social change in the larger world around them. The blame for the destruction of forest resources is often laid on shifting cultivators. But the forces bringing about the rapid loss of tropical forests at the end of the 20th century are the same forces that led to the destruction of the forests of Europe, urbanization, industrialization, increased affluence, populational growth and geographical expansion and the application the latest technology to extract ever more resources from the environment in pursuit of wealth and political power by competing groups. However we must know that those who practice Agriculture are at the receiving end of the social stratum.

Studies of small, isolated and pre-capitalist groups and their relationships with their environments suggests that the roots of the contemporary problem lie deep in human behavioral patterns, for even in these simple societies, competition and conflict can be identified as the main force driving them into contradiction with their environments.

Alternative Practice in the pre-Columbian Amazon basin

Slash-and-char, as opposed to slash-and-burn, may create self-perpetuating soil fertility that sup-ports sedentary agriculture, but the society so sustained may still be overturned, as above.

Organic Farming

Vegetables from organic farming.

Organic farming is an alternative agricultural system which originated early in the 20th century in reaction to rapidly changing farming practices. Organic agriculture continues to be developed by various organic agriculture organizations today. It relies on fertilizers of organic origin such as compost, manure, green manure, and bone meal and places emphasis on techniques such as crop rotation and companion planting. Biological pest control, mixed cropping and the fostering of insect predators are encouraged. In general, organic standards are designed to allow the use of naturally occurring substances while prohibiting or strictly limiting synthetic substances. For instance, naturally occurring pesticides such as pyrethrin and rotenone are permitted, while synthetic fertilizers and pesticides are generally prohibited. Synthetic substances that are allowed include, for example, copper sulfate, elemental sulfur and Ivermectin. Genetically modified organisms, nanomaterials, human sewage sludge, plant growth regulators, hormones, and antibiotic use in livestock husbandry are prohibited. Reasons for advocation of organic farming include real or perceived advantages in sustainability, openness, independence, health, food security, and food safety, although the match between perception and reality is continually challenged.

Organic agricultural methods are internationally regulated and legally enforced by many nations, based in large part on the standards set by the International Federation of Organic Agriculture Movements (IFOAM), an international umbrella organization for organic farming organizations established in 1972. Organic agriculture can be defined as:

an integrated farming system that strives for sustainability, the enhancement of soil fertility and biological diversity whilst, with rare exceptions, prohibiting synthetic pesticides, antibiotics, synthetic fertilizers, genetically modified organisms, and growth hormones.

Since 1990 the market for organic food and other products has grown rapidly, reaching $63 billion worldwide in 2012. This demand has driven a similar increase in organically managed farmland that grew from 2001 to 2011 at a compounding rate of 8.9% per annum. As of 2011, approximately 37,000,000 hectares (91,000,000 acres) worldwide were farmed organically, representing approximately 0.9 percent of total world farmland.

History

Agriculture was practiced for thousands of years without the use of artificial chemicals. Artificial fertilizers were first created during the mid-19th century. These early fertilizers were cheap, powerful, and easy to transport in bulk. Similar advances occurred in chemical pesticides in the 1940s, leading to the decade being referred to as the 'pesticide era'. These new agricultural techniques, while beneficial in the short term, had serious longer term side effects such as soil compaction, erosion, and declines in overall soil fertility, along with health concerns about toxic chemicals entering the food supply. In the late 1800s and early 1900s, soil biology scientists began to seek ways to remedy these side effects while still maintaining higher production.

Biodynamic agriculture was the first modern system of agriculture to focus exclusively on organic methods. Its development began in 1924 with a series of eight lectures on agriculture given by Rudolf Steiner. These lectures, the first known presentation of what later came to be known as organic agriculture, were held in response to a request by farmers who noticed degraded soil conditions and a deterioration in the health and quality of crops and livestock resulting from the use of chemical fertilizers. The one hundred eleven attendees, less than half of whom were farmers, came from six countries, primarily Germany and Poland. The lectures were published in November 1924; the first English translation appeared in 1928 as *The Agriculture Course.*

In 1921, Albert Howard and his wife Gabrielle Howard, accomplished botanists, founded an Institute of Plant Industry to improve traditional farming methods in India. Among other things, they brought improved implements and improved animal husbandry methods from their scientific training; then by incorporating aspects of the local traditional methods, developed protocalls for the rotation of crops, erosion prevention techniques, and the systematic use of composts and manures. Stimulated by these experiences of traditional farming, when Albert Howard returned to Britain in the early 1930s he began to promulgate a system of natural agriculture.

In July 1939, Ehrenfried Pfeiffer, the author of the standard work on biodynamic agriculture (*Bio-Dynamic Farming and Gardening*), came to the UK at the invitation of Walter James, 4th Baron Northbourne as a presenter at the Betteshanger Summer School and Conference on Biodynamic Farming at Northbourne's farm in Kent. One of the chief purposes of the conference was to bring together the proponents of various approaches to organic agriculture in order that they might cooperate within a larger movement. Howard attended the conference, where he met Pfeiffer. In the following year, Northbourne published his manifesto of organic farming, *Look to the Land*, in which he coined the term "organic farming." The Betteshanger conference has been described as the 'missing link' between biodynamic agriculture and other forms of organic farming.

In 1940 Howard published his *An Agricultural Testament.* In this book he adopted Northbourne's terminology of "organic farming." Howard's work spread widely, and he became known as the "father of organic farming" for his work in applying scientific knowledge and principles to various traditional and natural methods. In the United States J.I. Rodale, who was keenly interested both in Howard's ideas and in biodynamics, founded in the 1940s both a working organic farm for trials and experimentation, The Rodale Institute, and the Rodale Press to teach and advocate organic methods to the wider public. These became important influences on the spread of organic agriculture. Further work was done by Lady Eve Balfour in the United Kingdom, and many others across the world.

Increasing environmental awareness in the general population in modern times has transformed the originally supply-driven organic movement to a demand-driven one. Premium prices and some government subsidies attracted farmers. In the developing world, many producers farm according to traditional methods that are comparable to organic farming, but not certified, and that may not include the latest scientific advancements in organic agriculture. In other cases, farmers in the developing world have converted to modern organic methods for economic reasons.

Terminology

Biodynamic agriculturists, who based their work on Steiner's spiritually-oriented anthroposophy, used the term "organic" to indicate that a farm should be viewed as a living organism, in the sense of the following quotation:

"An organic farm, properly speaking, is not one that uses certain methods and substances and avoids others; it is a farm whose structure is formed in imitation of the structure of a natural system that has the integrity, the independence and the benign dependence of an organism"

— Wendell Berry, "The Gift of Good Land"

The use of "organic" popularized by Howard and Rodale, on the other hand, refers more narrowly to the use of organic matter derived from plant compost and animal manures to improve the humus content of soils, grounded in the work of early soil scientists who developed what was then called "humus farming." Since the early 1940s the two camps have tended to merge.

Methods

Organic cultivation of mixed vegetables in Capay, California. Note the hedgerow in the background.

"Organic agriculture is a production system that sustains the health of soils, ecosystems and people. It relies on ecological processes, biodiversity and cycles adapted to local conditions, rather than the use of inputs with adverse effects. Organic agriculture combines tradition, innovation and science to benefit the shared environment and promote fair relationships and a good quality of life for all involved..."

— *International Federation of Organic Agriculture Movements*

Organic farming methods combine scientific knowledge of ecology and modern technology with traditional farming practices based on naturally occurring biological processes. Organic farming methods are studied in the field of agroecology. While conventional agriculture uses synthetic pesticides and water-soluble synthetically purified fertilizers, organic farmers are restricted by regulations to using natural pesticides and fertilizers. An example of a natural pesticide is pyrethrin, which is found naturally in the Chrysanthemum flower. The principal methods of organic farming include crop rotation, green manures and compost, biological pest control, and mechanical cultivation. These measures use the natural environment to enhance agricultural productivity: legumes are planted to fix nitrogen into the soil, natural insect predators are encouraged, crops are rotated to confuse pests and renew soil, and natural materials such as potassium bicarbonate and mulches are used to control disease and weeds. Genetically modified seeds and animals are excluded.

While organic is fundamentally different from conventional because of the use of carbon based fertilizers compared with highly soluble synthetic based fertilizers and biological pest control instead of synthetic pesticides, organic farming and large-scale conventional farming are not entirely mutually exclusive. Many of the methods developed for organic agriculture have been borrowed by more conventional agriculture. For example, Integrated Pest Management is a multifaceted strategy that uses various organic methods of pest control whenever possible, but in conventional farming could include synthetic pesticides only as a last resort.

Crop Diversity

Organic farming encourages Crop diversity. The science of agroecology has revealed the benefits of polyculture (multiple crops in the same space), which is often employed in organic farming. Planting a variety of vegetable crops supports a wider range of beneficial insects, soil microorganisms, and other factors that add up to overall farm health. Crop diversity helps environments thrive and protects species from going extinct.

Soil Management

Organic farming relies heavily on the natural breakdown of organic matter, using techniques like green manure and composting, to replace nutrients taken from the soil by previous crops. This biological process, driven by microorganisms such as mycorrhiza, allows the natural production of nutrients in the soil throughout the growing season, and has been referred to as *feeding the soil to feed the plant*. Organic farming uses a variety of methods to improve soil fertility, including crop rotation, cover cropping, reduced tillage, and application of compost. By reducing tillage, soil is not inverted and exposed to air; less carbon is lost to the atmosphere resulting in more soil organic carbon. This has an added benefit of carbon sequestration, which can reduce green house gases and help reverse climate change.

Plants need nitrogen, phosphorus, and potassium, as well as micronutrients and symbiotic relationships with fungi and other organisms to flourish, but getting enough nitrogen, and particularly synchronization so that plants get enough nitrogen at the right time (when plants need it most), is a challenge for organic farmers. Crop rotation and green manure ("cover crops") help to provide nitrogen through legumes (more precisely, the *Fabaceae* family), which fix nitrogen from the atmosphere through symbiosis with rhizobial bacteria. Intercropping, which is sometimes used for insect and disease control, can also increase soil nutrients, but the competition between the legume and the crop can be problematic and wider spacing between crop rows is required. Crop residues can be ploughed back into the soil, and different plants leave different amounts of nitrogen, potentially aiding synchronization. Organic farmers also use animal manure, certain processed fertilizers such as seed meal and various mineral powders such as rock phosphate and green sand, a naturally occurring form of potash that provides potassium. Together these methods help to control erosion. In some cases pH may need to be amended. Natural pH amendments include lime and sulfur, but in the U.S. some compounds such as iron sulfate, aluminum sulfate, magnesium sulfate, and soluble boron products are allowed in organic farming.

Mixed farms with both livestock and crops can operate as ley farms, whereby the land gathers fertility through growing nitrogen-fixing forage grasses such as white clover or alfalfa and grows cash crops or cereals when fertility is established. Farms without livestock ("stockless") may find it more difficult to maintain soil fertility, and may rely more on external inputs such as imported manure as well as grain legumes and green manures, although grain legumes may fix limited nitrogen because they are harvested. Horticultural farms that grow fruits and vegetables in protected conditions often relay even more on external inputs.

Biological research into soil and soil organisms has proven beneficial to organic farming. Varieties of bacteria and fungi break down chemicals, plant matter and animal waste into productive soil nutrients. In turn, they produce benefits of healthier yields and more productive soil for future crops. Fields with less or no manure display significantly lower yields, due to decreased soil microbe community. Increased manure improves biological activity, providing a healthier, more arable soil system and higher yields.

Weed Management

Organic weed management promotes weed suppression, rather than weed elimination, by enhancing crop competition and phytotoxic effects on weeds. Organic farmers integrate cultural, biological, mechanical, physical and chemical tactics to manage weeds without synthetic herbicides.

Organic standards require rotation of annual crops, meaning that a single crop cannot be grown in the same location without a different, intervening crop. Organic crop rotations frequently include weed-suppressive cover crops and crops with dissimilar life cycles to discourage weeds associated with a particular crop. Research is ongoing to develop organic methods to promote the growth of natural microorganisms that suppress the growth or germination of common weeds.

Other cultural practices used to enhance crop competitiveness and reduce weed pressure include selection of competitive crop varieties, high-density planting, tight row spacing, and late planting into warm soil to encourage rapid crop germination.

Mechanical and physical weed control practices used on organic farms can be broadly grouped as:

- Tillage - Turning the soil between crops to incorporate crop residues and soil amendments; remove existing weed growth and prepare a seedbed for planting; turning soil after seeding to kill weeds, including cultivation of row crops;

- Mowing and cutting - Removing top growth of weeds;

- Flame weeding and thermal weeding - Using heat to kill weeds; and

- Mulching - Blocking weed emergence with organic materials, plastic films, or landscape fabric.

Some critics, citing work published in 1997 by David Pimentel of Cornell University, which described an epidemic of soil erosion worldwide, have raised concerned that tillage contribute to the erosion epidemic. The FAO and other organizations have advocated a 'no-till' approach to both conventional and organic farming, and point out in particular that crop rotation techniques used in organic farming are excellent no-till approaches. A study published in 2005 by Pimentel and colleagues confirmed that 'Crop rotations and cover cropping (green manure) typical of organic agriculture reduce soil erosion, pest problems, and pesticide use.' Some naturally sourced chemicals are allowed for herbicidal use. These include certain formulations of acetic acid (concentrated vinegar), corn gluten meal, and essential oils. A few selective bioherbicides based on fungal pathogens have also been developed. At this time, however, organic herbicides and bioherbicides play a minor role in the organic weed control toolbox.

Weeds can be controlled by grazing. For example, geese have been used successfully to weed a range of organic crops including cotton, strawberries, tobacco, and corn, reviving the practice of keeping cotton patch geese, common in the southern U.S. before the 1950s. Similarly, some rice farmers introduce ducks and fish to wet paddy fields to eat both weeds and insects.

Controlling Other Organisms

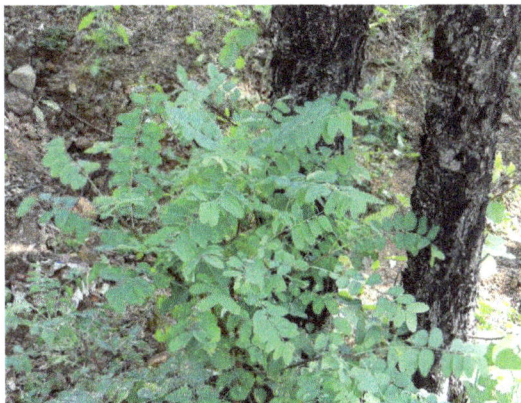

Chloroxylon is used for Pest Management in Organic Rice Cultivation in Chhattisgarh, India

Organisms aside from weeds that cause problems on organic farms include arthropods (e.g., insects, mites), nematodes, fungi and bacteria. Organic practices include, but are not limited to:

- encouraging predatory beneficial insects to control pests by serving them nursery plants and/or an alternative habitat, usually in a form of a shelterbelt, hedgerow, or beetle bank;

- encouraging beneficial microorganisms;

- rotating crops to different locations from year to year to interrupt pest reproduction cycles;

- planting companion crops and pest-repelling plants that discourage or divert pests;

- using row covers to protect crops during pest migration periods;

- using biologic pesticides and herbicides

- using stale seed beds to germinate and destroy weeds before planting

- using sanitation to remove pest habitat;

- Using insect traps to monitor and control insect populations.

- Using physical barriers, such as row covers

Examples of predatory beneficial insects include minute pirate bugs, big-eyed bugs, and to a lesser extent ladybugs (which tend to fly away), all of which eat a wide range of pests. Lacewings are also effective, but tend to fly away. Praying mantis tend to move more slowly and eat less heavily. Parasitoid wasps tend to be effective for their selected prey, but like all small insects can be less effective outdoors because the wind controls their movement. Predatory mites are effective for controlling other mites.

Naturally derived insecticides allowed for use on organic farms use include *Bacillus thuringiensis* (a bacterial toxin), pyrethrum (a chrysanthemum extract), spinosad (a bacterial metabolite), neem (a tree extract) and rotenone (a legume root extract). Fewer than 10% of organic farmers use these pesticides regularly; one survey found that only 5.3% of vegetable growers in California use rotenone while 1.7% use pyrethrum. These pesticides are not always more safe or environmentally friendly than synthetic pesticides and can cause harm. The main criterion for organic pesticides is that they are naturally derived, and some naturally derived substances have been controversial. Controversial natural pesticides include rotenone, copper, nicotine sulfate, and pyrethrums Rotenone and pyrethrum are particularly controversial because they work by attacking the nervous system, like most conventional insecticides. Rotenone is extremely toxic to fish and can induce symptoms resembling Parkinson's disease in mammals. Although pyrethrum (natural pyrethrins) is more effective against insects when used with piperonyl butoxide (which retards degradation of the pyrethrins), organic standards generally do not permit use of the latter substance.

Naturally derived fungicides allowed for use on organic farms include the bacteria *Bacillus subtilis* and *Bacillus pumilus*; and the fungus *Trichoderma harzianum*. These are mainly effective for diseases affecting roots. Compost tea contains a mix of beneficial microbes, which may attack or out-compete certain plant pathogens, but variability among formulations and preparation methods may contribute to inconsistent results or even dangerous growth of toxic microbes in compost teas.

Some naturally derived pesticides are not allowed for use on organic farms. These include nicotine sulfate, arsenic, and strychnine.

Synthetic pesticides allowed for use on organic farms include insecticidal soaps and horticultural oils for insect management; and Bordeaux mixture, copper hydroxide and sodium bicarbonate for managing fungi. Copper sulfate and Bordeaux mixture (copper sulfate plus lime), approved for organic use in various jurisdictions, can be more environmentally problematic than some synthetic fungicides dissallowed in organic farming Similar concerns apply to copper hydroxide. Repeated application of copper sulfate or copper hydroxide as a fungicide may eventually result in copper accumulation to toxic levels in soil, and admonitions to avoid excessive accumulations of copper in soil appear in various organic standards and elsewhere. Environmental concerns for several kinds of biota arise at average rates of use of such substances for some crops. In the European Union, where replacement of copper-based fungicides in organic agriculture is a policy priority, research is seeking alternatives for organic production.

Livestock

For livestock like these healthy cows vaccines play an important part in animal health since antibiotic therapy is prohibited in organic farming

Raising livestock and poultry, for meat, dairy and eggs, is another traditional farming activity that complements growing. Organic farms attempt to provide animals with natural living conditions and feed. Organic certification verifies that livestock are raised according to the USDA organic regulations throughout their lives. These regulations include the requirement that all animal feed must be certified organic.

Organic livestock may be, and must be, treated with medicine when they are sick, but drugs cannot be used to promote growth, their feed must be organic, and they must be pastured.

Also, horses and cattle were once a basic farm feature that provided labor, for hauling and plowing, fertility, through recycling of manure, and fuel, in the form of food for farmers and other animals. While today, small growing operations often do not include livestock, domesticated animals are a desirable part of the organic farming equation, especially for true sustainability, the ability of a farm to function as a self-renewing unit.

Genetic Modification

A key characteristic of organic farming is the rejection of genetically engineered plants and animals. On 19 October 1998, participants at IFOAM's 12th Scientific Conference issued the Mar del Plata Declaration, where more than 600 delegates from over 60 countries voted unanimously to exclude the use of genetically modified organisms in food production and agriculture.

Although opposition to the use of any transgenic technologies in organic farming is strong, agricultural researchers Luis Herrera-Estrella and Ariel Alvarez-Morales continue to advocate integration of transgenic technologies into organic farming as the optimal means to sustainable agriculture, particularly in the developing world, as does author and scientist Pamela Ronald, who views this kind of biotechnology as being consistent with organic principles.

Although GMOs are excluded from organic farming, there is concern that the pollen from genetically modified crops is increasingly penetrating organic and heirloom seed stocks, making it difficult, if not impossible, to keep these genomes from entering the organic food supply. Differing regulations among countries limits the availability of GMOs to certain countries, as described in the article on regulation of the release of genetic modified organisms.

Tools

Organic farmers use a number of traditional farm tools to do farming. Due to the goals of sustainability in organic farming, organic farmers try to minimize their reliance on fossil fuels. In the developing world on small organic farms tools are normally constrained to hand tools and diesel powered water pumps. Some organic farmers make use of renewable energy on the farm and can even make use of agrivoltaics or other onsite colocation of power production and agriculture. A recent study evaluated the use of open-source 3-D printers (called RepRaps using a bioplastic polylactic acid (PLA) on organic farms. PLA is a strong biodegradable and recyclable thermoplastic appropriate for a range of representative products in five categories of prints: handtools, food processing, animal management, water management and hydroponics. Such open source hardware is attractive to all types of small farmers as it provides control for farmers over their own equipment; this is exemplified by Open Source Ecology, Farm Hack and Farmbot.io.

Standards

Standards regulate production methods and in some cases final output for organic agriculture. Standards may be voluntary or legislated. As early as the 1970s private associations certified organic producers. In the 1980s, governments began to produce organic production guidelines. In the 1990s, a trend toward legislated standards began, most notably with the 1991 EU-Eco-regulation developed for European Union, which set standards for 12 countries, and a 1993 UK program. The EU's program was followed by a Japanese program in 2001, and in 2002 the U.S. created the National Organic Program (NOP). As of 2007 over 60 countries regulate organic farming (IFOAM 2007:11). In 2005 IFOAM created the Principles of Organic Agriculture, an international guideline for certification criteria. Typically the agencies accredit certification groups rather than individual farms.

Organic production materials used in and foods are tested independently by the Organic Materials Review Institute.

Composting

Using manure as a fertiliser risks contaminating food with animal gut bacteria, including pathogenic strains of E. coli that have caused fatal poisoning from eating organic food. To combat this risk, USDA organic standards require that manure must be sterilized through high temperature thermophilic composting. If raw animal manure is used, 120 days must pass before the crop is harvested if the final product comes into direct contact with the soil. For products that don't directly contact soil, 90 days must pass prior to harvest.

Economics

The economics of organic farming, a subfield of agricultural economics, encompasses the entire process and effects of organic farming in terms of human society, including social costs, opportunity costs, unintended consequences, information asymmetries, and economies of scale. Although the scope of economics is broad, agricultural economics tends to focus on maximizing yields and efficiency at the farm level. Economics takes an anthropocentric approach to the value of the natural world: biodiversity, for example, is considered beneficial only to the extent that it is valued by people and increases profits. Some entities such as the European Union subsidize organic farming, in large part because these countries want to account for the externalities of reduced water use, reduced water contamination, reduced soil erosion, reduced carbon emissions, increased biodiversity, and assorted other benefits that result from organic farming.

Traditional organic farming is labor and knowledge-intensive whereas conventional farming is capital-intensive, requiring more energy and manufactured inputs.

Organic farmers in California have cited marketing as their greatest obstacle.

Geographic Producer Distribution

The markets for organic products are strongest in North America and Europe, which as of 2001 are estimated to have $6 and $8 billion respectively of the $20 billion global market. As of 2007 Australasia has 39% of the total organic farmland, including Australia's 1,180,000 hectares (2,900,000 acres) but 97 percent of this land is sprawling rangeland (2007:35). US sales are 20x as much. Europe farms 23 percent of global organic farmland (6,900,000 ha (17,000,000 acres)), followed by Latin America with 19 percent (5.8 million hectares - 14.3 million acres). Asia has 9.5 percent while North America has 7.2 percent. Africa has 3 percent.

Besides Australia, the countries with the most organic farmland are Argentina (3.1 million hectares - 7.7 million acres), China (2.3 million hectares - 5.7 million acres), and the United States (1.6 million hectares - 4 million acres). Much of Argentina's organic farmland is pasture, like that of Australia (2007:42). Spain, Germany, Brazil (the world's largest agricultural exporter), Uruguay, and the UK follow the United States in the amount of organic land (2007:26).

In the European Union (EU25) 3.9% of the total utilized agricultural area was used for organic production in 2005. The countries with the highest proportion of organic land were Austria (11%) and Italy (8.4%), followed by the Czech Republic and Greece (both 7.2%). The lowest figures were shown for Malta (0.1%), Poland (0.6%) and Ireland (0.8%). In 2009, the proportion of organic land in the EU grew to 4.7%. The countries with highest share of agricultural land were Liechtenstein (26.9%), Austria (18.5%) and Sweden (12.6%). 16% of all farmers in Austria produced organically in 2010. By the same year the proportion of organic land increased to 20%.: In 2005 168,000 ha (415,000 ac) of land in Poland was under organic management. In 2012, 288,261 hectares (712,308 acres) were under organic production, and there were about 15,500 organic farmers; retail sales of organic products were EUR 80 million in 2011. As of 2012 organic exports were part of the government's economic development strategy.

After the collapse of the Soviet Union in 1991, agricultural inputs that had previously been purchased from Eastern bloc countries were no longer available in Cuba, and many Cuban farms converted to organic methods out of necessity. Consequently, organic agriculture is a mainstream practice in Cuba, while it remains an alternative practice in most other countries. Cuba's organic strategy includes development of genetically modified crops; specifically corn that is resistant to the palomilla moth

Growth

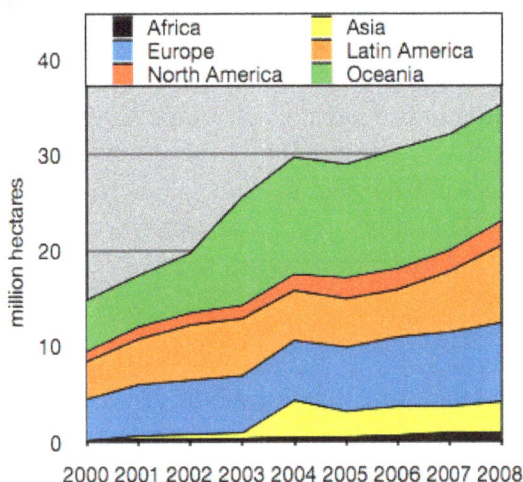

Organic farmland by world region (2000-2008)

In 2001, the global market value of certified organic products was estimated at USD $20 billion. By 2002, this was USD $23 billion and by 2007 more than USD $46 billion. By 2014, retail sales of organic products reached USD $80 billion worldwide. North America and Europe accounted for more than 90% of all organic product sales.

Organic agricultural land increased almost fourfold in 15 years, from 11 million hectares in 1999 to 43.7 million hectares in 2014. Between 2013 and 2014, organic agricultural land grew by 500,000 hectares worldwide, increasing in every region except Latin America. During this time period, Europe's organic farmland increased 260,000 hectares to 11.6 million total (+2.3%), Asia's increased 159,000 hectares to 3.6 million total (+4.7%), Africa's increased 54,000 hectares to 1.3 million total (+4.5%), and North America's increased 35,000 hectares to 3.1 million total (+1.1%). As of 2014, the country with the most organic land was Australia (17.2 million hectares), followed by Argentina (3.1 million hectares), and the United States (2.2 million hectares).

In 2013, the number of organic producers grew by almost 270,000, or more than 13%. By 2014, there were a reported 2.3 million organic producers in the world. Most of the total global increase took place in the Philippines, Peru, China, and Thailand. Overall, the majority of all organic producers are in India (650,000 in 2013), Uganda (190,552 in 2014), Mexico (169,703 in 2013) and the Philippines (165,974 in 2014).

Productivity

Studies comparing yields have had mixed results. These differences among findings can often be attributed to variations between study designs including differences in the crops studied and the methodology by which results were gathered.

A 2012 meta-analysis found that productivity is typically lower for organic farming than conventional farming, but that the size of the difference depends on context and in some cases may be very small. While organic yields can be lower than conventional yields, another meta-analysis published in Sustainable Agriculture Research in 2015, concluded that certain organic on-farm practices could help narrow this gap. Timely weed management and the application of manure in conjunction with legume forages/cover crops were shown to have positive results in increasing organic corn and soybean productivity. More experienced organic farmers were also found to have higher yields than other organic farmers who were just starting out.

Another meta-analysis published in the journal Agricultural Systems in 2011 analyzed 362 datasets and found that organic yields were on average 80% of conventional yields. The author's found that there are relative differences in this yield gap based on crop type with crops like soybeans and rice scoring higher than the 80% average and crops like wheat and potato scoring lower. Across global regions, Asia and Central Europe were found to have relatively higher yields and Northern Europe relatively lower than the average.

A 2007 study compiling research from 293 different comparisons into a single study to assess the overall efficiency of the two agricultural systems has concluded that "organic methods could produce enough food on a global per capita basis to sustain the current human population, and potentially an even larger population, without increasing the agricultural land base." The researchers also found that while in developed countries, organic systems on average produce

92% of the yield produced by conventional agriculture, organic systems produce 80% more than conventional farms in developing countries, because the materials needed for organic farming are more accessible than synthetic farming materials to farmers in some poor countries. This study was strongly contested by another study published in 2008, which stated, and was entitled, "Organic agriculture cannot feed the world" and said that the 2007 came up with "a major overestimation of the productivity of OA" "because data are misinterpreted and calculations accordingly are erroneous." Additional research needs to be conducted in the future to further clarify these claims.

Long Term Studies

A study published in 2005 compared conventional cropping, organic animal-based cropping, and organic legume-based cropping on a test farm at the Rodale Institute over 22 years. The study found that "the crop yields for corn and soybeans were similar in the organic animal, organic legume, and conventional farming systems". It also found that "significantly less fossil energy was expended to produce corn in the Rodale Institute's organic animal and organic legume systems than in the conventional production system. There was little difference in energy input between the different treatments for producing soybeans. In the organic systems, synthetic fertilizers and pesticides were generally not used". As of 2013 the Rodale study was ongoing and a thirty-year anniversary report was published by Rodale in 2012.

A long-term field study comparing organic/conventional agriculture carried out over 21 years in Switzerland concluded that "Crop yields of the organic systems averaged over 21 experimental years at 80% of the conventional ones. The fertilizer input, however, was 34 – 51% lower, indicating an efficient production. The organic farming systems used 20 – 56% less energy to produce a crop unit and per land area this difference was 36 – 53%. In spite of the considerably lower pesticide input the quality of organic products was hardly discernible from conventional analytically and even came off better in food preference trials and picture creating methods"

Profitability

In the United States, organic farming has been shown to be 2.9 to 3.8 times more profitable for the farmer than conventional farming when prevailing price premiums are taken into account. Globally, organic farming is between 22 and 35 percent more profitable for farmers than conventional methods, according to a 2015 meta-analysis of studies conducted across five continents.

The profitability of organic agriculture can be attributed to a number of factors. First, organic farmers do not rely on synthetic fertilizer and pesticide inputs, which can be costly. In addition, organic foods currently enjoy a price premium over conventionally produced foods, meaning that organic farmers can often get more for their yield.

The price premium for organic food is an important factor in the economic viability of organic farming. In 2013 there was a 100% price premium on organic vegetables and a 57% price premium for organic fruits. These percentages are based on wholesale fruit and vegetable prices, available through the United States Department of Agriculture's Economic Research Service. Price premiums exist not only for organic versus nonorganic crops, but may also vary depending on the venue where the product is sold: farmers markets, grocery stores, or wholesale to restaurants. For many producers, direct sales at farmers markets are most profitable because the farmer receives the entire markup, however this is also the most time and labor-intensive approach.

There have been signs of organic price premiums narrowing in recent years, which lowers the economic incentive for farmers to convert to or maintain organic production methods. Data from 22 years of experiments at the Rodale Institute found that, based on the current yields and production costs associated with organic farming in the United States, a price premium of only 10% is required to achieve parity with conventional farming. A separate study found that on a global scale, price premiums of only 5-7% percent were needed to break even with conventional methods. Without the price premium, profitability for farmers is mixed.

For markets and supermarkets organic food is profitable as well, and is generally sold at significantly higher prices than non-organic food.

Energy Efficiency

In the most recent assessments of the energy efficiency of organic versus conventional agriculture, results have been mixed regarding which form is more carbon efficient. Organic farm systems have more often than not been found to be more energy efficient, however, this is not always the case. More than anything, results tend to depend upon crop type and farm size.

A comprehensive comparison of energy efficiency in grain production, produce yield, and animal husbandry concluded that organic farming had a higher yield per unit of energy over the vast majority of the crops and livestock systems. For example, two studies - both comparing organically-versus conventionally-farmed apples - declare contradicting results, one saying organic farming is more energy efficient, the other saying conventionally is more efficient.

It has generally been found that the labor input per unit of yield was higher for organic systems compared with conventional production.

Sales and Marketing

Most sales are concentrated in developed nations. In 2008, 69% of Americans claimed to occasionally buy organic products, down from 73% in 2005. One theory for this change was that consumers were substituting "local" produce for "organic" produce.

Distributors

The USDA requires that distributors, manufacturers, and processors of organic products be certified by an accredited state or private agency. In 2007, there were 3,225 certified organic handlers, up from 2,790 in 2004.

Organic handlers are often small firms; 48% reported sales below $1 million annually, and 22% between $1 and $5 million per year. Smaller handlers are more likely to sell to independent natural grocery stores and natural product chains whereas large distributors more often market to natural product chains and conventional supermarkets, with a small group marketing to independent natural product stores. Some handlers work with conventional farmers to convert their land to organic with the knowledge that the farmer will have a secure sales outlet. This lowers the risk for the handler as well as the farmer. In 2004, 31% of handlers provided technical support on organic standards or production to their suppliers and 34% encouraged their suppliers to transition to organic. Smaller farms often join together in cooperatives to market their goods more effectively.

93% of organic sales are through conventional and natural food supermarkets and chains, while the remaining 7% of U.S. organic food sales occur through farmers' markets, foodservices, and other marketing channels.

Direct-to-Consumer Sales

In the 2012 Census, direct-to-consumer sales equaled $1.3 billion, up from $812 million in 2002, an increase of 60 percent. The number of farms that utilize direct-to-consumer sales was 144,530 in 2012 in comparison to 116,733 in 2002. Direct-to-consumer sales include farmers markets, community supported agriculture (CSA), on-farm stores, and roadside farm stands. Some organic farms also sell products direct to retailer, direct to restaurant and direct to institution. According to the 2008 Organic Production Survey, approximately 7% of organic farm sales went direct-to-consumers, 10% went direct to retailers, and approximately 83% went into wholesale markets. In comparison, only 0.4% of the value of convention agricultural commodities went direct-to-consumers.

While not all products sold at farmer's markets are certified organic, this direct-to-consumer avenue has become increasingly popular in local food distribution and has grown substantially since 1994. In 2014, there were 8,284 farmer's markets in comparison to 3,706 in 2004 and 1,755 in 1994, most of which are found in populated areas such as the Northeast, Midwest, and West Coast.

Labor and Employment

Organic production is more labor-intensive than conventional production. On the one hand, this increased labor cost is one factor that makes organic food more expensive. On the other hand, the increased need for labor may be seen as an "employment dividend" of organic farming, providing more jobs per unit area than conventional systems. The 2011 UNEP Green Economy Report suggests that "[a]n increase in investment in green agriculture is projected to lead to growth in employment of about 60 per cent compared with current levels" and that "green agriculture investments could create 47 million additional jobs compared with BAU2 over the next 40 years." The UNEP also argues that "[b]y greening agriculture and food distribution, more calories per person per day, more jobs and business opportunities especially in rural areas, and market-access opportunities, especially for developing countries, will be available."

World's Food Security

In 2007 the United Nations Food and Agriculture Organization (FAO) said that organic agriculture often leads to higher prices and hence a better income for farmers, so it should be promoted. However, FAO stressed that by organic farming one could not feed the current mankind, even less the bigger future population. Both data and models showed then that organic farming was far from sufficient. Therefore, chemical fertilizers were needed to avoid hunger. Other analysis by many agribusiness executives, agricultural and ecological scientists, and international agriculture experts revealed the opinion that organic farming would not only increase the world's food supply, but might be the only way to eradicate hunger.

FAO stressed that fertilizers and other chemical inputs can much increase the production, particularly in Africa where fertilizers are currently used 90% less than in Asia. For example, in Malawi the yield has been boosted using seeds and fertilizers. FAO also calls for using biotechnology, as it can help smallholder farmers to improve their income and food security.

Also NEPAD, development organization of African governments, announced that feeding Africans and preventing malnutrition requires fertilizers and enhanced seeds.

According to a more recent study in ScienceDigest, organic best management practices shows an average yield only 13% less than conventional. In the world's poorer nations where most of the world's hungry live, and where conventional agriculture's expensive inputs are not affordable by the majority of farmers, adopting organic management actually increases yields 93% on average, and could be an important part of increased food security.

Capacity Building in Developing Countries

Organic agriculture can contribute to ecologically sustainable, socio-economic development, especially in poorer countries. The application of organic principles enables employment of local resources (e.g., local seed varieties, manure, etc.) and therefore cost-effectiveness. Local and international markets for organic products show tremendous growth prospects and offer creative producers and exporters excellent opportunities to improve their income and living conditions.

Organic agriculture is knowledge intensive. Globally, capacity building efforts are underway, including localized training material, to limited effect. As of 2007, the International Federation of Organic Agriculture Movements hosted more than 170 free manuals and 75 training opportunities online.

In 2008 the United Nations Environmental Programme (UNEP) and the United Nations Conference on Trade and Development (UNCTAD) stated that "organic agriculture can be more conducive to food security in Africa than most conventional production systems, and that it is more likely to be sustainable in the long-term" and that "yields had more than doubled where organic, or near-organic practices had been used" and that soil fertility and drought resistance improved.

Millennium Development Goals

The value of organic agriculture (OA) in the achievement of the Millennium Development Goals (MDG), particularly in poverty reduction efforts in the face of climate change, is shown by its contribution to both income and non-income aspects of the MDGs. These benefits are expected to continue in the post-MDG era. A series of case studies conducted in selected areas in Asian countries by the Asian Development Bank Institute (ADBI) and published as a book compilation by ADB in Manila document these contributions to both income and non-income aspects of the MDGs. These include poverty alleviation by way of higher incomes, improved farmers' health owing to less chemical exposure, integration of sustainable principles into rural development policies, improvement of access to safe water and sanitation, and expansion of global partnership for development as small farmers are integrated in value chains.

A related ADBI study also sheds on the costs of OA programs and set them in the context of the costs of attaining the MDGs. The results show considerable variation across the case studies, suggesting that there is no clear structure to the costs of adopting OA. Costs depend on the efficiency of the OA adoption programs. The lowest cost programs were more than ten times less expensive than the highest cost ones. However, further analysis of the gains resulting from OA adoption reveals that the costs per person taken out of poverty was much lower than the estimates of the World Bank, based on income growth in general or based on the detailed costs of meeting some of the more quantifiable MDGs (e.g., education, health, and environment).

Externalities

Agriculture imposes negative externalities (uncompensated costs) upon society through land and other resource use, biodiversity loss, erosion, pesticides, nutrient runoff, water usage, subsidy payments and assorted other problems. Positive externalities include self-reliance, entrepreneurship, respect for nature, and air quality. Organic methods reduce some of these costs. In 2000 uncompensated costs for 1996 reached 2,343 million British pounds or £208 per ha (£84.20/ ac). A study of practices in the USA published in 2005 concluded that cropland costs the economy approximately 5 to 16 billion dollars ($30–96/ha - $12–39/ac), while livestock production costs 714 million dollars. Both studies recommended reducing externalities. The 2000 review included reported pesticide poisonings but did not include speculative chronic health effects of pesticides, and the 2004 review relied on a 1992 estimate of the total impact of pesticides.

It has been proposed that organic agriculture can reduce the level of some negative externalities from (conventional) agriculture. Whether the benefits are private or public depends upon the division of property rights.

Several surveys and studies have attempted to examine and compare conventional and organic systems of farming and have found that organic techniques, while not without harm, are less damaging than conventional ones because they reduce levels of biodiversity less than conventional systems do and use less energy and produce less waste when calculated per unit area.

A 2003 to 2005 investigation by the Cranfield University for the Department for Environment Food and Rural Affairs in the UK found that it is difficult to compare the Global Warming Potential (GWP), acidification and eutrophication emissions but "Organic production often results in increased burdens, from factors such as N leaching and N2O emissions", even though primary energy use was less for most organic products. N2O is always the largest GWP contributor except in tomatoes. However, "organic tomatoes always incur more burdens (except pesticide use)". Some emissions were lower "per area", but organic farming always required 65 to 200% more field area than non-organic farming. The numbers were highest for bread wheat (200+ % more) and potatoes (160% more).

The situation was shown dramatically in a comparison of a modern dairy farm in Wisconsin with one in New Zealand in which the animals grazed extensively. Using total farm emissions per kg milk produced as a parameter, the researchers showed that production of methane from belching was higher in the New Zealand farm, while carbon dioxide production was higher in the Wisconsin farm. Output of nitrous oxide, a gas with an estimated global warming potential 310 times that of carbon dioxide was also higher in the New Zealand farm. Methane from manure handling was similar in the two types of farm. The explanation for the finding relates to the different diets used on these farms, being based more completely on forage (and hence more fibrous) in New Zealand and containing less concentrate than in Wisconsin. Fibrous diets promote a higher proportion of acetate in the gut of ruminant animals, resulting in a higher production of methane that must be released by belching. When cattle are given a diet containing some concentrates (such as corn and soybean meal) in addition to grass and silage, the pattern of ruminal fermentation alters from acetate to mainly propionate. As a result, methane production is reduced. Capper et al. compared the environmental impact of US dairy production in 1944 and 2007. They calculated that the carbon "footprint" per billion kg (2.2 billion lb) of milk produced in 2007 was 37 percent that of equivalent milk production in 1944.

Environmental Impact and Emissions

Researchers at Oxford university analyzed 71 peer-reviewed studies and observed that organic products are sometimes worse for the environment. Organic milk, cereals, and pork generated higher greenhouse gas emissions per product than conventional ones but organic beef and olives had lower emissions in most studies. Usually organic products required less energy, but more land. Per unit of product, organic produce generates higher nitrogen leaching, nitrous oxide emissions, ammonia emissions, eutrophication and acidification potential than when conventionally grown. Other differences were not significant. The researchers concluded "Most of the studies that compared biodiversity in organic and conventional farming demonstrated lower environmental impacts from organic farming." The researchers believe that the ideal outcome would be to develop new systems that consider both the environment, including setting land aside for wildlife and sustainable forestry, and the development of ways to produce the highest yields possible using both conventional and organic methods.

Proponents of organic farming have claimed that organic agriculture emphasizes closed nutrient cycles, biodiversity, and effective soil management providing the capacity to mitigate and even reverse the effects of climate change and that organic agriculture can decrease fossil fuel emissions. "The carbon sequestration efficiency of organic systems in temperate climates is almost double (575-700 kg carbon per ha per year - 510-625 lb/ac/an) that of conventional treatment of soils, mainly owing to the use of grass clovers for feed and of cover crops in organic rotations."

Critics of organic farming methods believe that the increased land needed to farm organic food could potentially destroy the rainforests and wipe out many ecosystems.

Nutrient Leaching

According to the meta-analysis of 71 studies, nitrogen leaching, nitrous oxide emissions, ammonia emissions, eutrophication potential and acidification potential were higher for organic products, although in one study "nitrate leaching was 4.4-5.6 times higher in conventional plots than organic plots".

Excess nutrients in lakes, rivers, and groundwater can cause algal blooms, eutrophication, and subsequent dead zones. In addition, nitrates are harmful to aquatic organisms by themselves.

Land Use

The Oxford meta-analysis of 71 studies proved that organic farming requires 84% more land, mainly due to lack of nutrients but sometimes due to weeds, diseases or pests, lower yielding animals and land required for fertility building crops. While organic farming does not necessarily save land for wildlife habitats and forestry in all cases, the most modern breakthroughs in organic are addressing these issues with success.

Professor Wolfgang Branscheid says that organic animal production is not good for the environment, because organic chicken requires doubly as much land as "conventional" chicken and organic pork a quarter more. According to a calculation by Hudson Institute, organic beef requires triply as much land. On the other hand, certain organic methods of animal husbandry have been shown to restore desertified, marginal, and/or otherwise unavailable land to agricultural productivity and wildlife. Or by getting both forage and cash crop production from the same fields simultaneously, reduce net land use.

In England organic farming yields 55% of normal yields. While in other regions of the world, organic methods have started producing record yields.

Pesticides

A sign outside of an organic apple orchard in Pateros, Washington reminding orchardists not to spray pesticides on these trees.

Food Quality and Safety

While there may be some differences in the amounts of nutrients and anti-nutrients when organically produced food and conventionally produced food are compared, the variable nature of food production and handling makes it difficult to generalize results, and there is insufficient evidence to make claims that organic food is safer or healthier than conventional food. Claims that organic food tastes better are not supported by evidence.

Soil Conservation

Supporters claim that organically managed soil has a higher quality and higher water retention. This may help increase yields for organic farms in drought years. Organic farming can build up soil organic matter better than conventional no-till farming, which suggests long-term yield benefits from organic farming. An 18-year study of organic methods on nutrient-depleted soil concluded that conventional methods were superior for soil fertility and yield for nutrient-depleted soils in cold-temperate climates, arguing that much of the benefit from organic farming derives from imported materials that could not be regarded as self-sustaining.

In *Dirt: The Erosion of Civilizations*, geomorphologist David Montgomery outlines a coming crisis from soil erosion. Agriculture relies on roughly one meter of topsoil, and that is being depleted ten times faster than it is being replaced. No-till farming, which some claim depends upon pesticides, is one way to minimize erosion. However, a 2007 study by the USDA's Agricultural Research Service has found that manure applications in tilled organic farming are better at building up the soil than no-till.

Biodiversity

The conservation of natural resources and biodiversity is a core principle of organic production. Three broad management practices (prohibition/reduced use of chemical pesticides and inorganic fertilizers; sympathetic management of non-cropped habitats; and preservation of mixed farming) that are largely intrinsic (but not exclusive) to organic farming are particularly beneficial for farmland wildlife. Using practices that attract or introduce beneficial insects, provide habitat for birds and mammals, and provide conditions that increase soil biotic diversity serve to supply vital ecological services to organic production systems. Advantages to certified organic operations that implement these types of production practices include: 1) decreased dependence on outside fertility inputs; 2) reduced pest management costs; 3) more reliable sources of clean water; and 4) better pollination.

Nearly all non-crop, naturally occurring species observed in comparative farm land practice studies show a preference for organic farming both by abundance and diversity. An average of 30% more species inhabit organic farms. Birds, butterflies, soil microbes, beetles, earthworms, spiders, vegetation, and mammals are particularly affected. Lack of herbicides and pesticides improve biodiversity fitness and population density. Many weed species attract beneficial insects that improve soil qualities and forage on weed pests. Soil-bound organisms often benefit because of increased bacteria populations due to natural fertilizer such as manure, while experiencing reduced intake of herbicides and pesticides. Increased biodiversity, especially from beneficial soil microbes and mycorrhizae have been proposed as an explanation for the high yields experienced by some organic plots, especially in light of the differences seen in a 21-year comparison of organic and control fields.

Biodiversity from organic farming provides capital to humans. Species found in organic farms enhance sustainability by reducing human input (e.g., fertilizers, pesticides).

The USDA's Agricultural Marketing Service (AMS) published a *Federal Register* notice on 15 January 2016, announcing the National Organic Program (NOP) final guidance on Natural Resources and Biodiversity Conservation for Certified Organic Operations. Given the broad scope of natural resources which includes soil, water, wetland, woodland and wildlife, the guidance provides examples of practices that support the underlying conservation principles and demonstrate compliance with USDA organic regulations § 205.200. The final guidance provides organic certifiers and farms with examples of production practices that support conservation principles and comply with the USDA organic regulations, which require operations to maintain or improve natural resources. The final guidance also clarifies the role of certified operations (to submit an OSP to a certifier), certifiers (ensure that the OSP describes or lists practices that explain the operator's monitoring plan and practices to support natural resources and biodiversity conservation), and inspectors (onsite inspection) in the implementation and verification of these production practices.

A wide range of organisms benefit from organic farming, but it is unclear whether organic methods confer greater benefits than conventional integrated agri-environmental programs. Organic farming is often presented as a more biodiversity-friendly practice, but the generality of the beneficial effects of organic farming is debated as the effects appear often species- and context-dependent, and current research has highlighted the need to quantify the relative effects of local- and landscape-scale management on farmland biodiversity. There are four key issues when comparing the impacts on biodiversity of organic and conventional farming: (1) It remains unclear whether a holistic whole-farm approach (i.e. organic) provides greater benefits to biodiversity than carefully targeted prescriptions applied to relatively small areas of cropped and/or non-cropped habitats within conventional agriculture (i.e. agri-environment schemes); (2) Many comparative studies encounter methodological problems, limiting their ability to draw quantitative conclusions; (3) Our knowledge of the impacts of organic farming in pastoral and upland agriculture is limited; (4) There remains a pressing need for longitudinal, system-level studies in order to address these issues and to fill in the gaps in our knowledge of the impacts of organic farming, before a full appraisal of its potential role in biodiversity conservation in agroecosystems can be made.

Regional Support for Organic Farming

India

In India, states such as Sikkim and Kerala have planned to shift to fully organic cultivation by 2015 and 2016 respectively.

References

- Horne, Paul Anthony (2008). Integrated pest management for crops and pastures. CSIRO Publishing. p. 2. ISBN 978-0-643-09257-0.

- Stinner, D.H (2007). "The Science of Organic Farming". In William Lockeretz. Organic Farming: An International History. Oxfordshire, UK & Cambridge, Massachusetts: CAB International (CABI). ISBN 978-1-84593-289-3. Retrieved 30 April 2013.

- Vogt G (2007). Lockeretz W, ed. Chapter 1: The Origins of Organic Farming. Organic Farming: An International History. CABI Publishing. pp. 9–30. ISBN 9780851998336.

- Szykitka, Walter (2004). The Big Book of Self-Reliant Living: Advice and Information on Just About Everything You Need to Know to Live on Planet Earth. Globe-Pequot. p. 343. ISBN 978-1-59228-043-8.

- Pamela Ronald; Raoul Admachak (April 2008). "Tomorrow's Table: Organic Farming, Genetics and the Future of Food". Oxford University Press. ISBN 0195301757.

- Halberg, Niels (2006). Global development of organic agriculture: challenges and prospects. CABI. p. 297. ISBN 978-1-84593-078-3.

- Blair, Robert. (2012). Organic Production and Food Quality: A Down to Earth Analysis. Wiley-Blackwell, Oxford, UK. ISBN 978-0-8138-1217-5

- "USDSA Guidance Natural Resources and Biodiversity Conservation"(PDF). Agricultural Marketing Service, National Organic Program. United States Department of Agriculture. 15 January 2016. Retrieved 5 March2016

- Henckel, Laura (20 May 2015). "Organic fields sustain weed metacommunity dynamics in farmland landscapes". Proceedings of the Royal Society B. Retrieved 28 February 2016.

- Greene, Catherine. "USDA Economic Research Service - Organic Prices". www.ers.usda.gov. Retrieved 25 March 2016.

- "2012 Census Drilldown: Organic and Local Food | National Sustainable Agriculture Coalition". sustainableagriculture.net. Retrieved 2016-04-19.

- Low, Sarah (November 2011). "Direct and Intermediated Marketing of Local Foods in the United States" (PDF). USDA. Retrieved 2016-04-18.

- Suzie Key; Julian K-C Ma & Pascal MW Drake (1 June 2008). "Genetically modified plants and human health". Journal of the Royal Society of Medecine. pp. 290–298. Retrieved 11 March 2015.

Tillage: A Comprehensive Study

Tillage is one of the oldest methods of agriculture that is still in practice. Tillage involves the preparation of the soil before planting. Techniques of tillage alter according to the nature of the soil, its depth and drainage. Some techniques such as no-till farming, strip till farming and mulch-till farming are also explained in this chapter.

Tillage

Cultivating after an early rain

Tillage is the agricultural preparation of soil by mechanical agitation of various types, such as digging, stirring, and overturning. Examples of human-powered tilling methods using hand tools include shovelling, picking, mattock work, hoeing, and raking. Examples of draft-animal-powered or mechanized work include ploughing (overturning with moldboards or chiseling with chisel shanks), rototilling, rolling with cultipackers or other rollers, harrowing, and cultivating with cultivator shanks (teeth). Small-scale gardening and farming, for household food production or small business production, tends to use the smaller-scale methods above, whereas medium- to large-scale farming tends to use the larger-scale methods. There is a fluid continuum, however. Any type of gardening or farming, but especially larger-scale commercial types, may also use low-till or no-till methods as well.

Tillage is often classified into two types, primary and secondary. There is no strict boundary between them so much as a loose distinction between tillage that is deeper and more thorough (primary) and tillage that is shallower and sometimes more selective of location (secondary).

Primary tillage such as ploughing tends to produce a rough surface finish, whereas secondary tillage tends to produce a smoother surface finish, such as that required to make a good seedbed for many crops. Harrowing and rototilling often combine primary and secondary tillage into one operation.

"Tillage" can also mean the land that is tilled. The word "cultivation" has several senses that overlap substantially with those of "tillage". In a general context, both can refer to agriculture. Within agriculture, both can refer to any of the kinds of soil agitation described above. Additionally, "cultivation" or "cultivating" may refer to an even narrower sense of shallow, selective secondary tillage of row crop fields that kills weeds while sparing the crop plants.

Tillage Systems

Reduced Tillage

Plough tilling the field

Reduced tillage leaves between 15 and 30% residue cover on the soil or 500 to 1000 pounds per acre (560 to 1100 kg/ha) of small grain residue during the critical erosion period. This may involve the use of a chisel plow, field cultivators, or other implements.

Intensive Tillage

Intensive tillage leaves less than 15% crop residue cover or less than 500 pounds per acre (560 kg/ha) of small grain residue. This type of tillage is often referred to as conventional tillage but as conservational tillage is now more widely used than intensive tillage (in the United States), it is often not appropriate to refer to this type of tillage as conventional. Intensive tillage often involves multiple operations with implements such as a mold board, disk, and/or chisel plow. Then a finisher with a harrow, rolling basket, and cutter can be used to prepare the seed bed. There are many variations.

Conservation Tillage

Conservation tillage leaves at least 30% of crop residue on the soil surface, or at least 1,000 lb/ac (1,100 kg/ha) of small grain residue on the surface during the critical soil erosion period. This slows water movement, which reduces the amount of soil erosion. Conservation tillage also benefits farmers by reducing fuel consumption and soil compaction. By reducing the number of times the farmer travels over the field, farmers realize significant savings in fuel and labor. In most years since 1997, conservation tillage was used in US cropland more than intensive or reduced tillage.

However, conservation tillage delays warming of the soil due to the reduction of dark earth exposure to the warmth of the spring sun, thus delaying the planting of the next year's spring crop of corn.

- No-till - Never use a plow, disk, etc. ever again. Aims for 100% ground cover.

- Strip-Till - Narrow strips are tilled where seeds will be planted, leaving the soil in between the rows untilled.

- Mulch-till

- Rotational Tillage - Tilling the soil every two years or less often (every other year, or every third year, etc.).

- Ridge-Till

- Zone tillage - This form of conservation tillage is further explained below.

Zone Tillage

Zone tillage is a form of modified deep tillage in which only narrow strips are tilled, leaving soil in between the rows untilled. This type of tillage agitates the soil to help reduce soil compaction problems and to improve internal soil drainage.

Purpose

Zone tillage is designed to only disrupt the soil in a narrow strip directly below the crop row. In comparison to no-till, which relies on the previous year's plant residue to protect the soil and aides in postponement of the warming of the soil and crop growth in Northern climates, zone tillage creates approximately a 5-inch-wide strip that simultaneously breaks up plow pans, assists in warming the soil and helps to prepare a seedbed. When combined with cover crops, zone tillage helps replace lost organic matter, slows the deterioration of the soil, improves soil drainage, increases soil water and nutrient holding capacity, and allows necessary soil organisms to survive.

Usage

It has been successfully used on farms in the mid-west and west for over 40 years and is currently used on more than 36% of the U.S. farmland. Some specific states where zone tillage is currently in practice are Pennsylvania, Connecticut, Minnesota, Indiana, Wisconsin, and Illinois.

Unfortunately, there aren't consistent yield results in the Northern Cornbelt states; however, there is still interest in deep tillage within the agriculture industry. In areas that are not well-drained, deep tillage may be used as an alternative to installing more expensive tile drainage.

Effects of Tillage

Positive

Plowing:

- Loosens and aerates the top layer of soil, which facilitates planting the crop
- Helps mix harvest residue, organic matter (humus), and nutrients evenly into the soil
- Mechanically destroys weeds
- Dries the soil before seeding (in wetter climates tillage aids in keeping the soil drier)
- When done in autumn, helps exposed soil crumble over winter through frosting and de-frosting, which helps prepare a smooth surface for spring planting

Negative

- Dries the soil before seeding
- Soil loses a lot of nutrients, like nitrogen and fertilizer, and its ability to store water
- Decreases the water infiltration rate of soil. (Results in more runoff and erosion since the soil absorbs water slower than before)
- Tilling the soil results in dislodging the cohesiveness of the soil particles thereby inducing erosion.
- Chemical runoff
- Reduces organic matter in the soil
- Reduces microbes, earthworms, ants, etc.
- Destroys soil aggregates
- Compaction of the soil, also known as a tillage pan
- Eutrophication (nutrient runoff into a body of water)
- Can attract slugs, cut worms, army worms, and harmful insects to the left over residues.
- Crop diseases can be harbored in surface residues

General Comments

- The type of implement makes the most difference, although other factors can have an effect.
- Tilling in absolute darkness (night tillage) might reduce the number of weeds that sprout following the tilling operation by half. Light is necessary to break the dormancy of some

weed species' seed, so if fewer seeds are exposed to light during the tilling process, fewer will sprout. This may help reduce the amount of herbicides needed for weed control.

- Greater speeds, when using certain tillage implements (disks and chisel plows), lead to more intensive tillage (i.e., less residue is on the soil surface).

- Increasing the angle of disks causes residues to be buried more deeply. Increasing their concavity makes them more aggressive.

- Chisel plows can have spikes or sweeps. Spikes are more aggressive.

- Percentage residue is used to compare tillage systems because the amount of crop residue affects the soil loss due to erosion.

- See Soybean management practices to see what types of tillage are currently recommended for Soybean Production.

Definitions

Primary tillage loosens the soil and mixes in fertilizer and/or plant material, resulting in soil with a rough texture.

Secondary tillage produces finer soil and sometimes shapes the rows, preparing the seed bed. It also provides weed control throughout the growing season during the maturation of the crop plants, unless such weed control is instead achieved with low-till or no-till methods involving herbicides.

The seed bed preparation can be done with harrows (of which there are many types and subtypes), dibbles, hoes, shovels, rotary tillers, subsoilers, ridge- or bed-forming tillers, rollers, or cultivators.

The weed control, to the extent that it is done via tillage, is usually achieved with cultivators or hoes, which disturb the top few centimeters of soil around the crop plants but with minimal disturbance of the crop plants themselves. The tillage kills the weeds via 2 mechanisms: uprooting them, burying their leaves (cutting off their photosynthesis), or a combination of both. Weed control both prevents the crop plants from being outcompeted by the weeds (for water and sunlight) and prevents the weeds from reaching their seed stage, thus reducing future weed population aggressiveness.

History of Tilling

Tilling with Hungarian Grey cattles

Tilling was first performed via human labor, sometimes involving slaves. Hoofed animals could also be used to till soil via trampling. The wooden plow was then invented. It could be pulled by mule, ox, elephant, water buffalo, or similar sturdy animal. Horses are generally unsuitable, though breeds such as the scyne could work. The steel plow allowed farming in the American Midwest, where tough prairie grasses and rocks caused trouble. Soon after 1900, the farm tractor was introduced, which eventually made modern large-scale agriculture possible.

Alternatives to Tilling

Modern agricultural science has greatly reduced the use of tillage. Crops can be grown for several years without any tillage through the use of herbicides to control weeds, crop varieties that tolerate packed soil, and equipment that can plant seeds or fumigate the soil without really digging it up. This practice, called no-till farming, reduces costs and environmental change by reducing soil erosion and diesel fuel usage.

Site Preparation of Forest Land

Site preparation is any of various treatments applied to a site in order to ready it for seeding or planting. The purpose is to facilitate the regeneration of that site by the chosen method. Site preparation may be designed to achieve, singly or in any combination: improved access, by reducing or rearranging slash, and amelioration of adverse forest floor, soil, vegetation, or other biotic factors. Site preparation is undertaken to ameliorate one or more constraints that would otherwise be likely to thwart the objectives of management. A valuable bibliography on the effects of soil temperature and site preparation on subalpine and boreal tree species has been prepared by McKinnon et al. (2002).

Site preparation is the work that is done before a forest area is regenerated. Some types of site preparation are burning.

Burning

Broadcast burning is commonly used to prepare clearcut sites for planting, e.g., in central British Columbia, and in the temperate region of North America generally.

Prescribed burning is carried out primarily for slash hazard reduction and to improve site conditions for regeneration; all or some of the following benefits may accrue:

a) Reduction of logging slash, plant competition, and humus prior to direct seeding, planting, scarifying or in anticipation of natural seeding in partially cut stands or in connection with seed-tree systems.

b) Reduction or elimination of unwanted forest cover prior to planting or seeding, or prior to preliminary scarification thereto.

c) Reduction of humus on cold, moist sites to favour regeneration.

d) Reduction or elimination of slash, grass, or brush fuels from strategic areas around forested land to reduce the chances of damage by wildfire.

Prescribed burning for preparing sites for direct seeding was tried on a few occasions in Ontario, but none of the burns was hot enough to produce a seedbed that was adequate without supplementary mechanical site preparation.

Changes in soil chemical properties associated with burning include significantly increased pH, which Macadam (1987) in the Sub-boreal Spruce Zone of central British Columbia found persisting more than a year after the burn. Average fuel consumption was 20 to 24 t/ha and the forest floor depth was reduced by 28% to 36%. The increases correlated well with the amounts of slash (both total and ≥7 cm diameter) consumed. The change in pH depends on the severity of the burn and the amount consumed; the increase can be as much as 2 units, a 100-fold change. Deficiencies of copper and iron in the foliage of white spruce on burned clearcuts in central British Columbia might be attributable to elevated pH levels.

Even a broadcast slash fire in a clearcut does not give a uniform burn over the whole area. Tarrant (1954), for instance, found only 4% of a 140-ha slash burn had burned severely, 47% had burned lightly, and 49% was unburned. Burning after windrowing obviously accentuates the subsequent heterogeneity.

Marked increases in exchangeable calcium also correlated with the amount of slash at least 7 cm in diameter consumed. Phosphorus availability also increased, both in the forest floor and in the 0 cm to 15 cm mineral soil layer, and the increase was still evident, albeit somewhat diminished, 21 months after burning. However, in another study in the same Sub-boreal Spruce Zone found that although it increased immediately after the burn, phosphorus availability had dropped to below pre-burn levels within 9 months.

Nitrogen will be lost from the site by burning, though concentrations in remaining forest floor were found by Macadam (1987) to have increased in 2 of 6 plots, the others showing decreases. Nutrient losses may be outweighed, at least in the short term, by improved soil microclimate through the reduced thickness of forest floor where low soil temperatures are a limiting factor.

The *Picea/Abies* forests of the Alberta foothills are often characterized by deep accumulations of organic matter on the soil surface and cold soil temperatures, both of which make reforestation difficult and result in a general deterioration in site productivity; Endean and Johnstone (1974) describe experiments to test prescribed burning as a means of seedbed preparation and site amelioration on representative clear-felled *Picea/Abies* areas. Results showed that, in general, prescribed burning did not reduce organic layers satisfactorily, nor did it increase soil temperature, on the sites tested. Increases in seedling establishment, survival, and growth on the burned sites were probably the result of slight reductions in the depth of the organic layer, minor increases in soil temperature, and marked improvements in the efficiency of the planting crews. Results also suggested that the process of site deterioration has not been reversed by the burning treatments applied.

Ameliorative Intervention

Slash weight (the oven-dry weight of the entire crown and that portion of the stem < 4 inches in diameter) and size distribution are major factors influencing the forest fire hazard on harvested sites. Forest managers interested in the application of prescribed burning for hazard reduction and silviculture, were shown a method for quantifying the slash load by Kiil (1968). In west-central Alberta, he felled, measured, and weighed 60 white spruce, graphed (a) slash weight per merchantable unit volume against diameter at breast height (dbh), and (b) weight of fine slash (<1.27 cm) also against dbh, and produced a table of slash weight and size distribution on one acre of a hypothetical stand of white spruce. When the diameter distribution of a stand is unknown, an estimate of slash weight and size distribution can be obtained from average stand diameter, number of trees per unit area, and merchantable cubic foot volume. The sample trees in Kiil's study had full symmetrical crowns. Densely growing trees with short and often irregular crowns would probably be overestimated; open-grown trees with long crowns would probably be underestimated.

The need to provide shade for young outplants of Engelmann spruce in the high Rocky Mountains is emphasized by the U.S. Forest Service. Acceptable planting spots are defined as microsites on the north and east sides of down logs, stumps, or slash, and lying in the shadow cast by such material. Where the objectives of management specify more uniform spacing, or higher densities, than obtainable from an existing distribution of shade-providing material, redistribution or importing of such material has been undertaken.

Access

Site preparation on some sites might be done simply to facilitate access by planters, or to improve access and increase the number or distribution of microsites suitable for planting or seeding.

Wang et al. (2000) determined field performance of white and black spruces 8 and 9 years after outplanting on boreal mixedwood sites following site preparation (Donaren disc trenching versus no trenching) in 2 plantation types (open versus sheltered) in southeastern Manitoba. Donaren trenching slightly reduced the mortality of black spruce but significantly increased the mortality of white spruce. Significant difference in height was found between open and sheltered plantations for black spruce but not for white spruce, and root collar diameter in sheltered plantations was significantly larger than in open plantations for black spruce but not for white spruce. Black spruce open plantation had significantly smaller volume (97 cm³) compared with black spruce sheltered (210 cm³), as well as white spruce open (175 cm³) and sheltered (229 cm³) plantations. White spruce open plantations also had smaller volume than white spruce sheltered plantations. For transplant stock, strip plantations had a significantly higher volume (329 cm³) than open plantations (204 cm³). Wang et al. (2000) recommended that sheltered plantation site preparation should be used.

Mechanical

Up to 1970, no "sophisticated" site preparation equipment had become operational in Ontario, but the need for more efficacious and versatile equipment was increasingly recognized. By this time, improvements were being made to equipment originally developed by field staff, and field testing of equipment from other sources was increasing.

According to J. Hall (1970), in Ontario at least, the most widely used site preparation technique was post-harvest mechanical scarification by equipment front-mounted on a bulldozer (blade, rake, V-plow, or teeth), or dragged behind a tractor (Imsett or S.F.I. scarifier, or rolling chopper). Drag type units designed and constructed by Ontario's Department of Lands and Forests used anchor chain or tractor pads separately or in combination, or were finned steel drums or barrels of various sizes and used in sets alone or combined with tractor pad or anchor chain units.

J. Hall's (1970) report on the state of site preparation in Ontario noted that blades and rakes were found to be well suited to post-cut scarification in tolerant hardwood stands for natural regeneration of yellow birch. Plows were most effective for treating dense brush prior to planting, often in conjunction with a planting machine. Scarifying teeth, e.g., Young's teeth, were sometimes used to prepare sites for planting, but their most effective use was found to be preparing sites for seeding, particularly in backlog areas carrying light brush and dense herbaceous growth. Rolling choppers found application in treating heavy brush but could be used only on stone-free soils. Finned drums were commonly used on jack pine–spruce cutovers on fresh brushy sites with a deep duff layer and heavy slash, and they needed to be teamed with a tractor pad unit to secure good distribution of the slash. The S.F.I. scarifier, after strengthening, had been "quite successful" for 2 years, promising trials were under way with the cone scarifier and barrel ring scarifier, and development had begun on a new flail scarifier for use on sites with shallow, rocky soils. Recognition of the need to become more effective and efficient in site preparation led the Ontario Department of Lands and Forests to adopt the policy of seeking and obtaining for field testing new equipment from Scandinavia and elsewhere that seemed to hold promise for Ontario conditions, primarily in the north. Thus, testing was begun of the Brackekultivator from Sweden and the Vako-Visko rotary furrower from Finland.

Mounding

Site preparation treatments that create raised planting spots have commonly improved outplant performance on sites subject to low soil temperature and excess soil moisture. Mounding can certainly have a big influence on soil temperature. Draper et al. (1985), for instance, documented this as well as the effect it had on root growth of outplants (Table 30).

The mounds warmed up quickest, and at soil depths of 0.5 cm and 10 cm averaged 10 and 7 °C higher, respectively, than in the control. On sunny days, daytime surface temperature maxima on the mound and organic mat reached 25 °C to 60 °C, depending on soil wetness and shading. Mounds reached mean soil temperatures of 10 °C at 10 cm depth 5 days after planting, but the control did not reach that temperature until 58 days after planting. During the first growing season, mounds had 3 times as many days with a mean soil temperature greater than 10 °C than did the control microsites.

Draper et al.'s (1985) mounds received 5 times the amount of photosynthetically active radiation (PAR) summed over all sampled microsites throughout the first growing season; the control treatment consistently received about 14% of daily background PAR, while mounds received over 70%. By November, fall frosts had reduced shading, eliminating the differential. Quite apart from its effect on temperature, incident radiation is also important photosynthetically. The average control microsite was exposed to levels of light above the compensation point for only 3 hours, i.e., one-quarter of the daily light period, whereas mounds received light above the compensation point for 11 hours, i.e., 86% of the same daily period. Assuming that incident light in the 100-600 $\mu Em^{-2}s^{-1}$ intensity range is the most important for photosynthesis, the mounds received over 4 times the total daily light energy that reached the control microsites.

Orientation of Linear Site Preparation, e.g., Disk-trenching

With linear site preparation, orientation is sometimes dictated by topography or other considerations, but the orientation can often be chosen. It can make a difference. A disk-trenching experiment in the Sub-boreal Spruce Zone in interior British Columbia investigated the effect on growth of young outplants (lodgepole pine) in 13 microsite planting positions: berm, hinge, and trench in each of north, south, east, and west aspects, as well as in untreated locations between the furrows. Tenth-year stem volumes of trees on south, east, and west-facing microsites were significantly greater than those of trees on north-facing and untreated microsites. However, planting spot selection was seen to be more important overall than trench orientation.

In a Minnesota study, the N–S strips accumulated more snow but snow melted faster than on E–W strips in the first year after felling. Snow-melt was faster on strips near the centre of the strip-felled area than on border strips adjoining the intact stand. The strips, 50 feet (15.24 m) wide, alternating with uncut strips 16 feet (4.88 m) wide, were felled in a *Pinus resinosa* stand, aged 90 to 100 years.

Classification of Tillage

No-till Farming

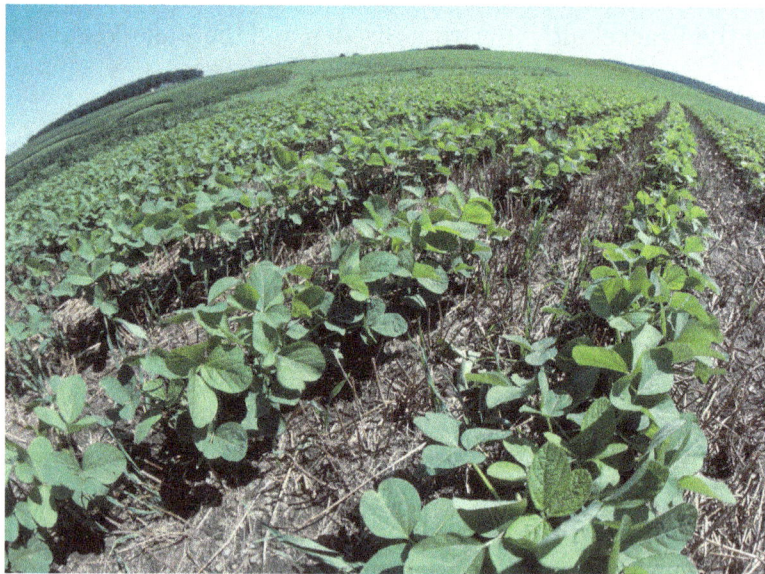

Young soybean plants thrive in and are protected by the residue of a wheat crop. This form of
no till farming provides good protection for the soil from erosion and helps retain moisture for the new crop.

No-till farming (also called zero tillage or direct drilling) is a way of growing crops or pasture from year to year without disturbing the soil through tillage. No-till is an agricultural technique which increases the amount of water that infiltrates into the soil and increases organic matter retention and cycling of nutrients in the soil. In many agricultural regions it can reduce or eliminate soil erosion. It increases the amount and variety of life in and on the soil, including disease-causing organisms and disease suppression organisms. The most powerful benefit of no-tillage is improvement in soil biological fertility, making soils more resilient. Farm operations are made much more efficient, particularly improved time of sowing and better trafficability of farm operations.

Background

Tilling is the process of removing plants or plant debris, usually for the purposes of planting more desirable species. This tilling can result in a flat seed bed or one that has formed areas, such as rows or raised beds, to enhance the growth of desired plants. It is an ancient technique with clear evidence of its use since at least 3000 B.C.

The effects of tillage can include soil compaction; loss of organic matter; degradation of soil aggregates; death or disruption of soil microbes and other organisms including mycorrhiza, arthropods, and earthworms; and soil erosion where topsoil is washed or blown away.

Origin of No-till for Modern Farms

The idea of modern no-till farming started in the 1940s with Edward H. Faulkner, author of *Plowman's Folly*, but it wasn't until the development of several chemicals after WWII that various researchers and farmers started to try out the idea. The first adopters of no-till include Klingman (North Carolina), Edward Faulkner, L.A. Porter (New Zealand), Harry and Lawrence Young (Herndon, Kentucky), the Instituto de Pesquisas Agropecuarias Meridional (1971 in Brazil) with Herbert Bartz.

Adoption Rate in the United States

No-till farming is widely used in the United States and the number of acres managed in this way continues to grow. This growth is supported by a decrease in costs related to tillage; no-till management results in fewer passes with equipment for approximately equal harvests, and the crop residue prevents evaporation of rainfall and increases water infiltration into the soil.

Issues

Profit, Economics, Yield

Studies have found that no-till farming can be more profitable if performed correctly.

Less tillage of the soil reduces labour, fuel, irrigation and machinery costs. No-till can increase yield because of higher water infiltration and storage capacity, and less erosion. Another benefit of no-till is that because of the higher water content, instead of leaving a field fallow it can make economic sense to plant another crop instead.

As sustainable agriculture becomes more popular, monetary grants and awards are becoming readily available to farmers who practice conservation tillage. Some large energy corporations which are among the greatest generators of fossil-fuel-related pollution may purchase carbon credits, which can encourage farmers to engage in conservation tillage. Under such schemes, the farmers' land is legally redefined as a carbon sink for the power generators' emissions. This helps the farmer in several ways, and it helps the energy companies meet regulatory demands for reduction of pollution, specifically carbon emissions.

No-till farming can increase organic (carbon based) matter in the soil, which is a form of carbon sequestration. However, there is debate over whether this increased sequestration detected in scientific studies of no-till agriculture is actually occurring, or is due to flawed testing methods or other factors. Regardless of this debate, there are still many other good reasons to use no-till, e.g. reduction in fossil fuel use, no erosion, soil quality.

Environmental

Carbon (air and soil) and Other Greenhouse Gases

No-till farming has carbon sequestration potential through storage of soil organic matter in the soil of crop fields. Whereas, when soil is tilled by machinery, the soil layers invert, air mixes in, and soil microbial activity dramatically increases over baseline levels. Tilling results in soil organic matter being broken down much more rapidly, and carbon is lost from the soil into the atmosphere. In addition to the effect on soil from tilling, emissions from the farm tractors increases carbon dioxide levels in the atmosphere.

Cropland soils are ideal for use as a carbon sink, since they have been depleted of carbon in most areas. It is estimated that 78 billion metric tonnes of carbon that was trapped in the soil has been released because of tillage. Conventional farming practices that rely on tillage have removed carbon from the soil ecosystem by removing crop residues such as left over corn stalks, and through the addition of chemical fertilizers which have the above-mentioned effects on soil microbes. By eliminating tillage, crop residues decompose where they lie, and by growing winter cover crops, carbon loss can be slowed and eventually reversed.

Nonetheless, a growing body of research is showing that no-till systems lose carbon stocks over time. Regarding a 2014 study of which he was principal investigator, University of Illinois soil scientist Ken Olson said this differing result occurs in part because tested soil samples need to include the full depth of rooting; 1–2 meters deep. He said, "That no-till subsurface layer is often losing more soil organic carbon stock over time than is gained in the surface layer." Also, there has not been a uniform definition of soil organic carbon sequestration among researchers. The study concludes, "Additional investments in SOC research is needed to better understand the agricultural management practices that are most likely to sequester SOC or at least retain more net SOC stocks."

In addition to keeping carbon in the soil, no-till farming reduces nitrous oxide (N_2O) emissions by 40-70%, depending on rotation. Nitrous oxide is a potent greenhouse gas, 300 times stronger than CO2, and stays in the atmosphere for 120 years. Fertilizing farmlands with (excessive) nitrogen increases the release of nitrous oxide.

Soil and Water

No-till farming improves soil quality (soil function), carbon, organic matter, aggregates, protecting the soil from erosion, evaporation of water, and structural breakdown. A reduction in tillage passes helps prevent the compaction of soil.

Recently, researchers at the Agricultural Research Service of the United States Department of Agriculture found that no-till farming makes soil much more stable than plowed soil. Their conclusions draw from over 19 years of collaborated tillage studies. No-till stores more carbon in the soil and carbon in the form of organic matter is a key factor in holding soil particles together. The first inch of no-till soil is two to seven times less vulnerable than that of plowed soil. The practice of no-till farming is especially beneficial to Great Plains farmers because of its resistance to erosion.

Crop residues left intact help both natural precipitation and irrigation water infiltrate the soil where it can be used. The crop residue left on the soil surface also limits evaporation, conserving water for plant growth. Soil compaction and no tillage-pan, soil absorbs more water and plants are able to grow their roots deeper into the soil and suck up more water.

Tilling a field reduces the amount of water, via evaporation, around 1/3 to 3/4 inches (0.85 to 1.9 cm) per pass. By no-tilling, this water stays in the soil, available to the plants.

Soil Biota, Wildlife, etc.

In no-till farming the soil is left intact and crop residue is left on the field. Therefore, soil layers, and in turn soil biota, are conserved in their natural state. No-tilled fields often have more beneficial insects and annelids, a higher microbial content, and a greater amount of soil organic material. Since there is no ploughing there is less airborne dust.

No-till farming increases the amount and variety of wildlife. This is the result of improved cover, reduced traffic and the reduced chance of destroying ground nesting birds and animals (plowing destroys all of them).

Albedo

Tillage lowers the albedo of croplands. The potential for global cooling as a result of increased Albedo in no till croplands is similar in magnitude to the biogeochemical (carbon sequestration) potential.

Historical Artifacts

Tilling regularly damages ancient structures under the soil such as long barrows. In the UK, half of the long barrows in Gloucestershire and almost all the burial mounds in Essex have been damaged. According to English Heritage modern tillage techniques have done as much damage in the last six decades as traditional tilling did in the previous six centuries. By using no-till methods these structures can be preserved and can be properly investigated instead of being destroyed.

Cost

Equipment

No-till farming requires specialized seeding equipment designed to plant seeds into undisturbed crop residues and soil. If the farmer has equipment designed for tillage farming, purchasing new equipment (seed drills for example) would be expensive and while the cost could be offset by selling off plows, etc. doing so is not usually done until the farmer decides to switch completely over (after trying it out for a few years). This would result in more money being invested into equipment in the short term (until old equipment is sold off).

Drainage

If a soil has poor drainage, it may need drainage tiles or other devices in order to help with the removal of excess water under no-till. Farmers should remember that water infiltration will improve after several years of a field being in no-till farming, so they may want to wait until 5–8 years have passed to see if the problems persists before deciding to invest in such an expensive system.

Gullies

Gullies can be a problem in the long-term. While much less soil is displaced by using no-till farming, any drainage gulleys that do form will get deeper each year since they aren't being smoothed out by plowing. This may necessitate either sod drainways, waterways, permanent drainways, cover crops, etc. Gully formation can be avoided entirely with proper water management practices, including the creation of swales on contour.

Increased Chemical Use

One of the purposes of tilling is to remove weeds. No-till farming does change weed composition drastically. Faster growing weeds may no longer be a problem in the face of increased competition, but shrubs and trees may begin to grow eventually.

Some farmers attack this problem with a "burn-down" herbicide such as glyphosate in lieu of tillage for seedbed preparation and because of this, no-till is often associated with increased chemical use in comparison to traditional tillage based methods of crop production. However, there are many agroecological alternatives to increased chemical use, such as winter cover crops and the mulch cover they provide, soil solarization or burning.

Management

No-till farming requires some different skills in order to do it successfully. As with any production system, if no-till isn't done correctly, yields can drop. A combination of technique, equipment, pesticides, crop rotation, fertilization, and irrigation have to be used for local conditions.

Cover Crops

Cover crops are used occasionally in no-till farming to help control weeds and increase nutrients in the soil (by using legumes) or by using plants with long roots to pull mobile nutrients back up to the surface from lower layers of the soil. Farmers experimenting with organic no-till use cover crops instead of tillage for controlling weeds, and are developing various methods to kill the cover crops (rollers, crimper, choppers, etc.) so that the newly planted crops can get enough light, water, nutrients, etc.

Disease, Pathogens, Insects and the use of Crop Rotations

With no-till farming, residue from the previous years crops lie on the surface of the field, cooling it and increasing the moisture. This can cause increased or decreased or variations of diseases that occur, but not necessarily at a higher or lower rate than conventional tillage. In order to help eliminate weed, pest and disease problems, crop rotations are used. By rotating the crops on a multi-year cycle, pests and diseases will decrease since the pests will no longer have a food supply to support their numbers.

Organic no-till Technique: The Cardboard Method

Some farmers who prefer to pursue a chemical-free management practice often rely on the use of normal, non-dyed corrugated cardboard for use on seed-beds and vegetable areas. Used correctly, cardboard placed on a specific area can

1. keep important fungal hyphae and microorganisms in the soil intact

2. prevent recurring weeds from popping up

3. increase residual nitrogen and plant nutrients by top-composting plant residues and

4. create valuable topsoil that is well suited for next years seeds or transplants.

The plant residues (left over plant matter originating from cover crops, grass clippings, original plant life etc.) will rot while underneath the cardboard so long as it remains sufficiently moist. This rotting attracts worms and other beneficial microorganisms to the site of decomposition, and over a series of a few seasons (usually Spring-->Fall or Fall-->Spring) and up to a few years, will create a layer of rich topsoil. Plants can then be direct seeded into the soil come spring, or holes can be cut into the cardboard to allow for transplantation. Using this method in conjunction with other sustainable practices such as composting/vermicompost, cover crops and rotations are often considered beneficial to both land and those who take from it.

Water Issues

No-till farming dramatically reduces the amount of erosion in a field. While much less soil is displaced, any gullies that do form will get deeper each year instead of being smoothed out by regular plowing. This may necessitate either sod drainways, waterways, permanent drainways, cover crops, etc.

A problem that occurs in some fields is water saturation in soils. Switching to no-till farming will help the drainage issue because of the qualities of soil under continuous no-till include a higher water infiltration rate.

Equipment

It is very important to have planting equipment that can properly penetrate through the residue, into the soil and prepare a good seedbed. Switching to no-till reduces the maximum amount of power needed from farm tractors, which means that a farmer can farm under no-till with a smaller tractor than if he/she were tilling. Using a smaller, lighter tractor has the added benefit of reducing compaction.

Soil Temperature

Another problem that growers face is that in the spring the soil will take longer to warm and dry, which may delay planting to a less ideal future date. One reason why the soil is slower to warm is that the field absorbs less solar energy as the residue covering the soil is a much lighter color than the black soil which would be exposed in conventional tillage. This can be managed by using row cleaners on a planter. Since the soil can be cooler, harvest can occur a few days later than a conventionally tilled field. Note: A cooler soil is also a benefit because water doesn't evaporate as fast.

Residue

On some crops, like continuous no-till corn, the thickness of the residue on the surface of the field can become a problem without proper preparation and/or equipment.

Fertilizer

One of the most common yield reducers is nitrogen being immobilized in the crop residue, which can take a few months to several years to decompose, depending on the crop's C to N ratio and the local environment. Fertilizer needs to be applied at a higher rate during the transition period while the soil rebuilds its organic matter. The nutrients in the organic matter will be eventually released back into the soil, so this is only a concern during the transition time frame (4–5 years for Kansas, USA). An innovative solution to this problem is to integrate animal husbandry in various ways to aid in the decomposition cycle.

Misconceptions

Need to Fluff the Soil

Although no-till farming often causes a slight increase in soil bulk density, periodic tilling is not needed to "fluff" the soil back up. No-till farming mimics the natural conditions under which most soils formed more closely than any other method of farming, in that the soil is left undisturbed except to place seeds in a position to germinate.

External Links

- No-Till Farmer Website

Similar Terms

No-till farming is not equivalent to conservation tillage or strip tillage. Conservation tillage is a group of practices that reduce the amount of tillage needed. No-till and strip tillage are both forms of conservation tillage. No-till is the practice of never tilling a field. Tilling every other year is called rotational tillage.

Strip-till

Strip-till is a conservation system that uses a minimum tillage. It combines the soil drying and warming benefits of conventional tillage with the soil-protecting advantages of no-till by disturbing only the portion of the soil that is to contain the seed row. This type of tillage is performed with special equipment and can require the farmer to make multiple trips, depending on the strip-till implement used, and field conditions. Each row that has been strip-tilled is usually about eight to ten inches wide.

A field planted using strip-till. Notice the crop residue of prior crop between the growing crop rows.

One of multiple variations of strip tillage equipment

Challenges of Both Strip-till and No-till Systems

In reduced tillage strategies, weed suppression can be difficult. In place of cultivation, a farmer can suppress weeds by managing a cover crop, mowing, crimping, or herbicide application. The purchase of mowing and crimping implements may represent an unjust expenditure. Additionally, finding an appropriate cover crop mix for adequate weed suppression may be difficult. Also, without mowing or crimping implements it may not be possible to achieve a kill on the cover crop. If mowing, crimping, and suppression with a cover crop mixture fail, herbicides can be applied. However, this may represent an increase in total farm expenses due to herbicides being used in place of cultivation for weed suppression.

There are some disadvantages specific to strip-till systems. Some farmers may not be able to strip-till if there is an early freeze. Though strip tillage can be successful without a global position system (GPS) based guidance, it can be beneficial. Lastly, strip-till systems requires a high-horsepower tractor; however, the energy requirement is less than with conventional tillage systems.

Differences in the Equipment Used

No-till planters have a disk opener and/or coulter that is located in front of the planting unit. This coulter is designed to cut through crop residue and into the hard crust of the soil. After the coulter has broken through the residue and crust, the disk opener of the planting unit slices the soil and the seed is dropped into the furrow that has been created and then a press wheel closes the furrow.

With strip-tillage systems more precision is needed. At the same time the field is strip-tilled, the fertilizer or chemical may be applied. If the meter of the chemical or fertilizer applicator is off slightly, an accurate rate may not be applied. This could result in increased expenses or reduction of the efficacy of the fertilizer or chemical program.

Effects on the Soils Properties

When oxygen is introduced into the soil via tillage, the decomposition of organic matter is accelerated. Carbon, nitrogen, and phosphorus were all higher in the no-till system than on reduced till, and conventional till systems in an Australian study.

Strip tillage has some similarities with no-till systems because the surface is protected with residue. However, strip-till also has a similar effect on soil properties as conventional tillage systems because the farmer still breaks the soil's crust which allows aerobic conditions to speed the decay of organic matter. A two-year study found that strip-till did not affect the amount of soil organic carbon or its extractable phosphorus.

Impacts on Productivity

In one study, yields were higher in the strip-tilled area than in the area where no-till was practiced. In a low phosphorus site, yield was 43.5 bu/a in strip-till compared to 41.5 bu/a in a no-till system. Yield is comparable to that of intensive tillage systems — without the cost.

Benefits of Strip Till

Strip till warms the soil, it allows an aerobic condition, and it allows for a better seedbed than no-till. Strip-till allows the soil's nutrients to be better adapted to the plant's needs, while still giving residue cover to the soil between the rows. The system will still allow for some soil water contact that could cause erosion, however, the amount of erosion on a strip-tilled field would be light compared to the amount of erosion on an intensively tilled field. Furthermore, when liquid fertilizer is being applied, it can be directly applied in these rows where the seed is being planted, reducing the amount of fertilizer needed while improving proximity of the fertilizer to the roots. Compared to intensive tillage, strip tillage saves considerable time and money. Strip tillage can reduce the amount of trips through a field down to two or possibly one trip when using a strip till implement combined with other machinery such as a planter, fertilizer spreader, and chemical sprayer. This can save the farmer a considerable amount of time and fuel, while reducing soil compaction due to few passes in a field. With the use of GPS-guided tractors, this precision farming can increase overall yields. Strip-till conserves more soil moisture compared to intensive tillage systems. However, compared to no-till, strip-till may in some cases reduce soil moisture.

Mulch-till

In agriculture mulch tillage or mulch-till fall under the umbrella term of conservation tillage in the United States and refer to seeding methods where a hundred percent of the soil surface is disturbed by tillage whereby crop residues are mixed with the soil and a certain amount of residues remain on the soil surface. A great variety of cultivator implements are used to perform mulch-till.

Mulch is material to regulate heat. This is done by covering it with any material like rice, straw or leaves or food waste.

References

- Olson K.R., Al-Kaisi M.M., Lal R., Lowery B. (2014). Experimental Consideration, Treatments, and Methods in Determining Soil Organic Carbon Sequestration Rates. Soil Sci. Soc. Am. J. 78:2:pp.348-360.

- Tallman, Susan. "No-Till Case Study, Richter Farm: Cover Crop Cocktails in a Forage-Based System". National Sustainable Agriculture Information Service. NCAT-ATTRA. Retrieved 8 April 2013.

- "CONSERVATION PRACTICE DEFINITIONS AND THEIR CORRESPONDING POTENTIAL TO ADVERSELY IMPACT CULTURAL RESOURCES" (PDF). Natural Resources Conservation Service. U.S. Department of Agriculture. p. 11. Retrieved 9 November 2013.

- "Indiana Job Sheet (340)" (docx) (ver. 1.3 ed.). Section with the heading "Operation and Maintenance" and "Termination": U.S. Department of Agriculture. October 2011. Retrieved 9 November 2013.

- "CONSERVATION TILLAGE IN THE UNITED STATES: AN OVERVIEW". okstate.edu. Institute of Agriculture and Natural Resources, University of Nebraska – Lincoln U.S.A. p. Figure 2. Retrieved 8 July 2013.

- Ray Hilborn. "Soils in Agriculture" (PPT--available as non-PPT by searching the path through a search engine). University of Washington. Retrieved 2013-08-28.

- Mahdi Al-Kaisi; Mark Hanna; Michael Tidman (13 May 2002). "Methods for measuring crop residue". Iowa State University. Retrieved 2012-12-28.

- D. B. Lobell, G. Bala and P. B. Duffy; D. B. Lobell; G. Bala; P. B. Duffy (2006-03-23). "Biogeophysical impacts of cropland management changes on climate" (PDF). 33. GEOPHYSICAL RESEARCH LETTERS. Retrieved 2012-07-02.

- Lal, Rattan. "No-Till Farming Offers A Quick Fix To Help Ward Off Host Of Global Problems". Researchnews. osu.edu. Retrieved 2010-05-09.

- D. B. Warburton and W. D. Klimstra; D. B. Warburton; W. D. Klimstra (1984-09-01). "Wildlife use of no-till and conventionally tilled corn fields". 39 (5). Journal of Soil and Water Conservation. Retrieved 2010-05-09.

Agricultural Productivity and Its Sources

Commercialisation of agriculture has led to great progress in the fields of plant modification, high-yielding seeds and agricultural tools. Agriculture is an important part of the economy and is still witnessing vast amounts of progress and development. Some themes in this chapter include the green revolution, extensive farming and intensive farming.

Agricultural Productivity

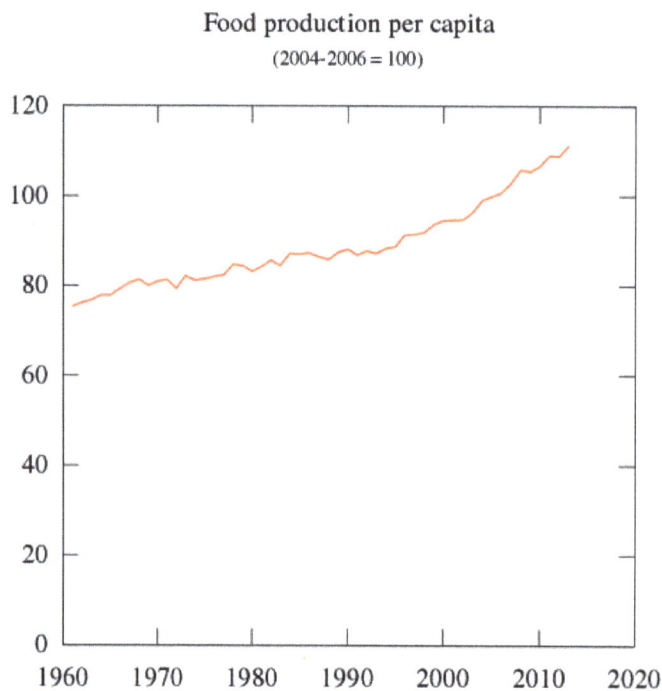

Food production per capita
(2004-2006 = 100)

Food production per capita since 1961

Agricultural productivity is measured as the ratio of agricultural outputs to agricultural inputs. While individual products are usually measured by weight, their varying densities make measuring overall agricultural output difficult. Therefore, output is usually measured as the market value of final output, which excludes intermediate products such as corn feed used in the meat industry. This output value may be compared to many different types of inputs such as labour and land (yield). These are called partial measures of productivity.

Grain production facilities

Agricultural productivity may also be measured by what is termed total factor productivity (TFP). This method of calculating agricultural productivity compares an index of agricultural inputs to an index of outputs. This measure of agricultural productivity was established to remedy the short-comings of the partial measures of productivity; notably that it is often hard to identify the factors cause them to change. Changes in TFP are usually attributed to technological improvements.

Sources of Agricultural Productivity

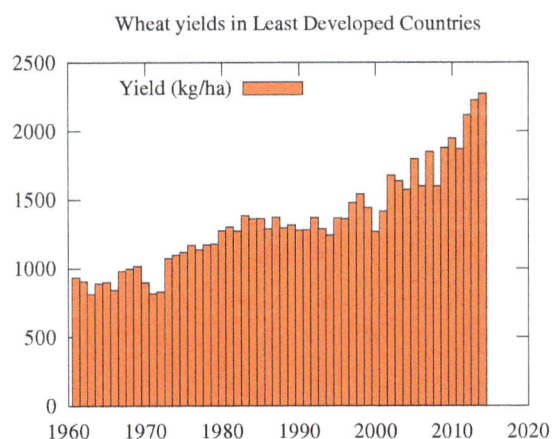

Wheat yields in least developed countries since 1961. The steep rise in crop yields in the U.S. began in the 1940s. The percentage of growth was fastest in the early rapid growth stage. In developing countries maize yields are still rapidly rising.

Some sources of agricultural productivity are:

- Mechanization

- High yield varieties, which were the basis of the Green revolution

- Fertilizers: Primary plant nutrients: nitrogen, phosphorus and potassium and secondary nutrients such as sulfur, zinc, copper, manganese, calcium, magnesium and molybdenum on deficient soil

- Liming of acid soils to raise pH and to provide calcium and magnesium

- Irrigation

- Herbicides

- Pesticides

- Increased plant density

- Animal feed made more digestible by processing

- Keeping animals indoors in cold weather

Importance of Agricultural Productivity

The productivity of a region's farms is important for many reasons. Aside from providing more food, increasing the productivity of farms affects the region's prospects for growth and competitiveness on the agricultural market, income distribution and savings, and labour migration. An increase in a region's agricultural productivity implies a more efficient distribution of scarce resources. As farmers adopt new techniques and differences, the more productive farmers benefit from an increase in their welfare while farmers who are not productive enough will exit the market to seek success elsewhere.

A cooperative dairy factory in Victoria.

As a region's farms become more productive, its comparative advantage in agricultural products increases, which means that it can produce these products at a lower opportunity cost than can other regions. Therefore, the region becomes more competitive on the world market, which means that it can attract more consumers since they are able to buy more of the products offered for the same amount of money.

Increases in agricultural productivity lead also to agricultural growth and can help to alleviate poverty in poor and developing countries, where agriculture often employs the greatest portion of the population. As farms become more productive, the wages earned by those who work in agriculture increase. At the same time, food prices decrease and food supplies become more stable. Labourers therefore have more money to spend on food as well as other products. This also leads to agricultural growth. People see that there is a greater opportunity to earn their living by farming and are attracted to agriculture either as owners of farms themselves or as labourers.

A liquid manure spreader.

However, it is not only the people employed in agriculture who benefit from increases in agricultural productivity. Those employed in other sectors also enjoy lower food prices and a more stable food supply. Their wages may also increase.

Agricultural productivity is becoming increasingly important as the world population continues to grow. India, one of the world's most populous countries, has taken steps in the past decades to increase its land productivity. Forty years ago, North India produced only wheat, but with the advent of the earlier maturing high-yielding wheats and rices, the wheat could be harvested in time to plant rice. This wheat/rice combination is now widely used throughout the Punjab, Haryana, and parts of Uttar Pradesh. The wheat yield of three tons and rice yield of two tons combine for five tons of grain per hectare, helping to feed India's 1.1 billion people.

Agricultural Productivity and Sustainable Development

Increase in agricultural productivity are often linked with questions about sustainability and sustainable development. Changes in agricultural practices necessarily bring changes in demands on resources. This means that as regions implement measures to increase the productivity of their farm land, they must also find ways to ensure that future generations will also have the resources they will need to live and thrive.

U.S. Agriculture Productivity

Between 1950 and 2000, during the so-called "second agricultural revolution of modern times", U.S. agricultural productivity rose fast, especially due to the development of new technologies. For example, the average amount of milk produced per cow increased from 5,314 pounds to 18,201 pounds per year (+242%), the average yield of corn rose from 39 bushels to 153 bushels per acre (+292%), and each farmer in 2000 produced on average 12 times as much farm output per hour worked as a farmer did in 1950.

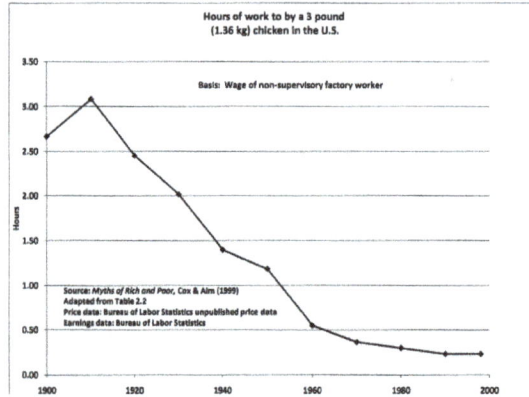

Hours of work to by a 3 pound
(1.36 kg) chicken in the U.S.

Basis: Wage of non-supervisory factory worker

Source: *Myths of Rich and Poor*, Cox & Alm (1999)
Adapted from Table 2.2
Price data: Bureau of Labor Statistics unpublished price data
Earnings data: Bureau of Labor Statistics

An hour's work in 1998 bought 11 times as much chicken as in 1900. Shown this way
includes the increase in overall buying power plus lowered costs from the farm to the consumer.

Productive Farms

Dairy cattle in Maryland

For many farmers (especially in non-industrial countries) agricultural productivity may mean much more. A productive farm is one that provides most of the resources necessary for the farmer's family to live, such as food, fuel, fiber, healing plants, etc. It is a farm which ensures food security as well as a way to sustain the well-being of a community. This implies that a productive farm is also one which is able to ensure proper management of natural resources, such as biodiversity, soil, water, etc. For most farmers, a productive farm would also produce more goods than required for the community in order to allow trade.

Some essential food products including bread, rice and pasta

Diversity in agricultural production is one key to productivity, as it enables risk management and preserves potentials for adaptation and change. Monoculture is an example of such a nondiverse production system. In a monocultural system a farmer may produce only crops, but no livestock, or only livestock and no crop.

The benefits of raising livestock, among others, are that it provides multiple goods, such as food, wool, hides, and transportation. It also has an important value in term of social relationships (such as gifts in weddings). In case of famine, when crops are not sufficient to ensure food safety, livestock can be used as food. Livestock may also provide manure, which can be used to fertilize cultivated soils, which increases soil productivity. On the other hand, in an agricultural system based only on raising livestock, food has to be bought from other farmers, and wastes produced cannot be easily disposed of. Production has many functions, and diversity is the foundation of such production. To ignore the complex functions provided by a farm is thought by many to turn agricultural production into a commodity.

Soil Fertility

Soil fertility refers to the ability of a soil to sustain plant growth, i.e. to provide plant habitat and result in sustained and consistent yields of high quality. A fertile soil has the following properties:

- The ability to supply essential plant nutrients and soil water in adequate amounts and proportions for plant growth and reproduction; and

- The absence of toxic substances which may inhibit plant growth.

The following properties contribute to soil fertility in most situations:

- Sufficient soil depth for adequate root growth and water retention;

- Good internal drainage, allowing sufficient aeration for optimal root growth (although some plants, such as rice, tolerate waterlogging);

- Topsoil with sufficient soil organic matter for healthy soil structure and soil moisture retention;

- Soil pH in the range 5.5 to 7.0 (suitable for most plants but some prefer or tolerate more acid or alkaline conditions);

- Adequate concentrations of essential plant nutrients in plant-available forms;

- Presence of a range of microorganisms that support plant growth.

In lands used for agriculture and other human activities, maintenance of soil fertility typically requires the use of soil conservation practices. This is because soil erosion and other forms of soil degradation generally result in a decline in quality with respect to one or more of the aspects indicated above.

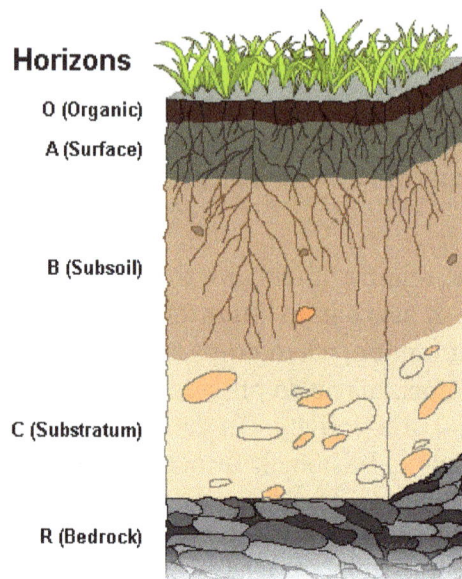

Horizons

O (Organic)
A (Surface)
B (Subsoil)
C (Substratum)
R (Bedrock)

Soil scientists use the capital letters O, A, B, C, and E to identify the master horizons, and lowercase letters for distinctions of these horizons. Most soils have three major horizons—the surface horizon (A), the subsoil (B), and the substratum (C). Some soils have an organic horizon (O) on the surface, but this horizon can also be buried. The master horizon, E, is used for subsurface horizons that have a significant loss of minerals (eluviation). Hard bedrock, which is not soil, uses the letter R.

Soil Fertilization

Bioavailable phosphorus is the element in soil that is most often lacking. Nitrogen and potassium are also needed in substantial amounts. For this reason these three elements are always identified on a commercial fertilizer analysis. For example, a 10-10-15 fertilizer has 10 percent nitrogen, 10 percent (P_2O_5) available phosphorus and 15 percent (K_2O) water-soluble potassium. Sulfur is the fourth element that may be identified in a commercial analysis—e.g. 21-0-0-24 which would contain 21% nitrogen and 24% sulfate.

Inorganic fertilizers are generally less expensive and have higher concentrations of nutrients than organic fertilizers. Also, since nitrogen, phosphorus and potassium generally must be in the inorganic forms to be taken up by plants, inorganic fertilizers are generally immediately bioavailable to plants without modification. However, some have criticized the use of inorganic fertilizers, claiming that the water-soluble nitrogen doesn't provide for the long-term needs of the plant and creates water pollution. Slow-release fertilizers may reduce leaching loss of nutrients and may make the nutrients that they provide available over a longer period of time.

Soil fertility is a complex process that involves the constant cycling of nutrients between organic and inorganic forms. As plant material and animal wastes are decomposed by micro-organisms, they release inorganic nutrients to the soil solution, a process referred to as mineralization. Those nutrients may then undergo further transformations which may be aided or enabled by soil micro-organisms. Like plants, many micro-organisms require or preferentially use inorganic forms of nitrogen, phosphorus or potassium and will compete with plants for these nutrients, tying up the nutrients in microbial biomass, a process often called immobilization. The balance between immobilization and mineralization processes depends on the balance and availability of major nutrients and organic carbon to soil microorganisms. Natural processes such as lightning strikes may fix atmospheric nitrogen by converting it to (NO_2). Denitrification may occur under anaerobic conditions (flooding) in the presence of denitrifying bacteria. The cations, primarily phosphate and potash, as well as many micronutrients are held in relatively strong bonds with the negatively charged portions of the soil in a process known as cation-exchange capacity.

In 2008 the cost of phosphorus as fertilizer more than doubled, while the price of rock phosphate as base commodity rose eight-fold. Recently the term peak phosphorus has been coined, due to the limited occurrence of rock phosphate in the world.

Light and CO_2 Limitations

Photosynthesis is the process whereby plants use light energy to drive chemical reactions which convert CO_2 into sugars. As such, all plants require access to both light and carbon dioxide to produce energy, grow and reproduce.

While typically limited by nitrogen, phosphorus and potassium, low levels of carbon dioxide can also act as a limiting factor on plant growth. Peer-reviewed and published scientific studies have shown that increasing CO_2 is highly effective at promoting plant growth up to levels over 300 ppm. Further increases in CO_2 can, to a very small degree, continue to increase net photosynthetic output.

Since higher levels of CO_2 have only a minimal impact on photosynthetic output at present levels (presently around 400 ppm and increasing), we should not consider plant growth to be limited by carbon dioxide. Other biochemical limitations, such as soil organic content, nitrogen in the soil, phosphorus and potassium, are far more often in short supply. As such, neither commercial nor scientific communities look to air fertilization as an effective or economic method of increasing production in agriculture or natural ecosystems. Furthermore, since microbial decomposition occurs faster under warmer temperatures, higher levels of CO_2 (which is one of the causes of unusually fast climate change) should be expected to increase the rate at which nutrients are leached out of soils and may have a negative impact on soil fertility.

Soil Depletion

Soil depletion occurs when the components which contribute to fertility are removed and not replaced, and the conditions which support soil's fertility are not maintained. This leads to poor crop yields. In agriculture, depletion can be due to excessively intense cultivation and inadequate soil management.

Soil fertility can be severely challenged when land use changes rapidly. For example, in Colonial New England, colonists made a number of decisions that depleted the soils, including: allowing herd animals to wander freely, not replenishing soils with manure, and a sequence of events that led to erosion. William Cronon wrote that "...the long-term effect was to put those soils in jeopardy. The removal of the forest, the increase in destructive floods, the soil compaction and close-cropping wrought by grazing animals, plowing--all served to increase erosion."

One of the most widespread occurrences of soil depletion as of 2008 is in tropical zones where nutrient content of soils is low. The combined effects of growing population densities, large-scale industrial logging, slash-and-burn agriculture and ranching, and other factors, have in some places depleted soils through rapid and almost total nutrient removal.

Topsoil depletion occurs when the nutrient-rich organic topsoil, which takes hundreds to thousands of years to build up under natural conditions, is eroded or depleted of its original organic material. Historically, many past civilizations' collapses can be attributed to the depletion of the topsoil. Since the beginning of agricultural production in the Great Plains of North America in the 1880s, about one-half of its topsoil has disappeared.

Depletion may occur through a variety of other effects, including overtillage (which damages soil structure), underuse of nutrient inputs which leads to mining of the soil nutrient bank, and salinization of soil.

Irrigation Water Effects

The quality of irrigation water is very important to maintain soil fertility and tilth, and for using more soil depth by the plants. When soil is irrigated with high alkaline water, unwanted sodium salts build up in the soil which would make soil draining capacity very poor. So plant roots can not penetrate deep in to the soil for optimum growth in Alkali soils. When soil is irrigated with low pH / acidic water, the useful salts (Ca, Mg, K, P, S, etc.) are removed by draining water from the acidic soil and in addition unwanted aluminium and manganese salts to the plants are dissolved from the soil impeding plant growth. When soil is irrigated with high salinity water or sufficient water is not draining out from the irrigated soil, the soil would convert in to saline soil or lose its fertility. Saline water enhance the turgor pressure or osmotic pressure requirement which impedes the off take of water and nutrients by the plant roots.

Top soil loss takes place in alkali soils due to erosion by rain water surface flows or drainage as they form colloids (fine mud) in contact with water. Plants absorb water-soluble inorganic salts only from the soil for their growth. Soil as such does not lose fertility just by growing crops but it lose its fertility due to accumulation of unwanted and depletion of wanted inorganic salts from the soil by improper irrigation and acid rain water (quantity and quality of water). The fertility of many soils which are not suitable for plant growth can be enhanced many times gradually by providing adequate irrigation water of suitable quality and good drainage from the soil.

Global Distribution

Global Soil Regions

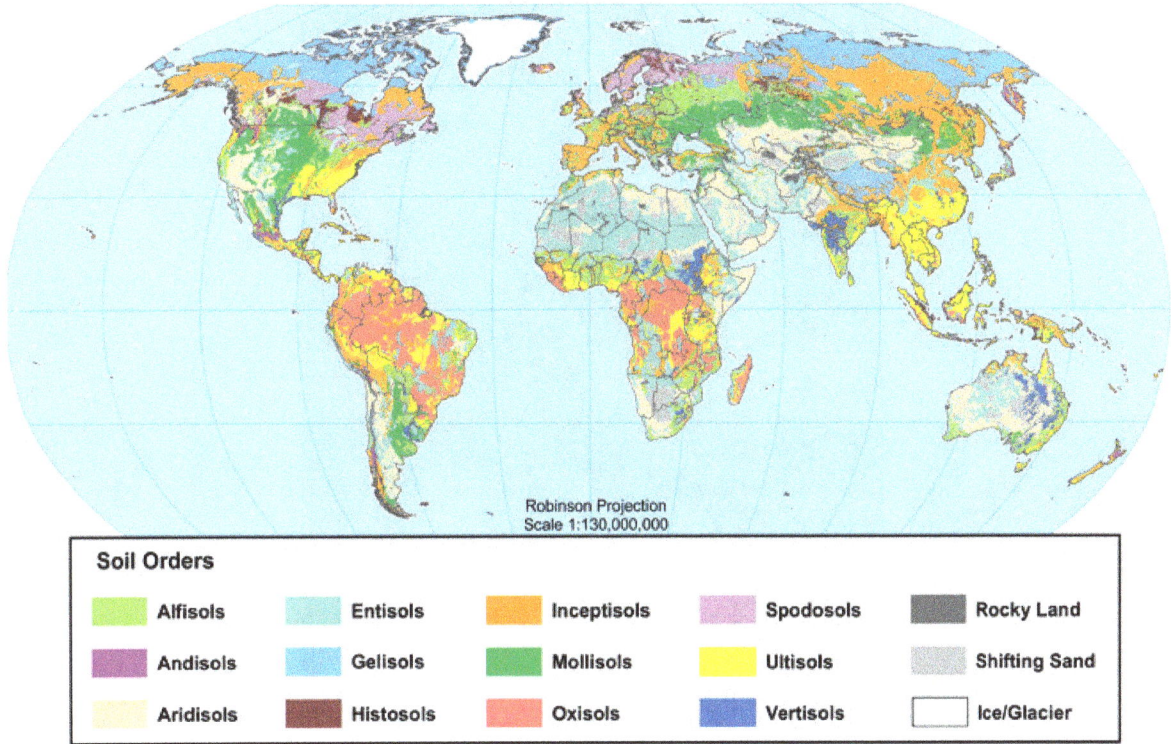

Soil Orders

Alfisols	Entisols	Inceptisols	Spodosols	Rocky Land
Andisols	Gelisols	Mollisols	Ultisols	Shifting Sand
Aridisols	Histosols	Oxisols	Vertisols	Ice/Glacier

USDA — NRCS — US Department of Agriculture / Natural Resources / Conservation Service — Soil Survey Division / World Soil Resources / soils.usda.gov/use/worldsoils — November 2005

Global distribution of soil types of the USDA soil taxonomy system. Mollisols, shown here in dark green, are a good (though not the only) indicator of high soil fertility. They coincide to a large extent with the world's major grain producing areas like the North American Prairie States, the Pampa and Gran Chaco of South America and the Ukraine-to-Central Asia Black Earth belt.

Extensive Farming

Extensive farming or extensive agriculture (as opposed to intensive farming) is an agricultural production system that uses small inputs of labor, fertilizers, and capital, relative to the land area being farmed.

Extensive farming most commonly refers to sheep and cattle farming in areas with low agricultural productivity, but can also refer to large-scale growing of wheat, barley, cooking oils and other grain crops in areas like the Murray-Darling Basin. Here, owing to the extreme age and poverty of the soils, yields per hectare are very low, but the flat terrain and very large farm sizes mean yields per unit of labour are high. Nomadic herding is an extreme example of extensive farming, where herders move their animals to use feed from occasional rainfalls.

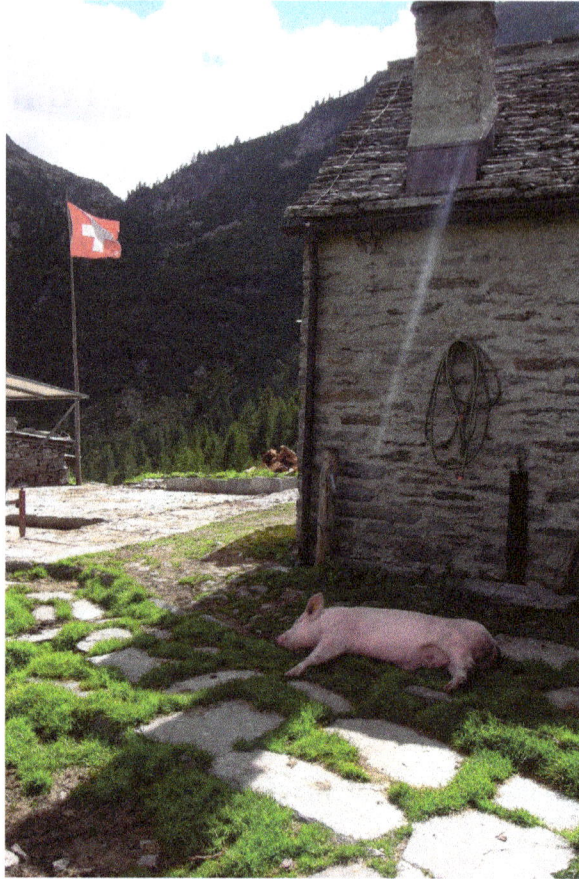

A small farm in the Swiss mountains. The land here is mostly rock and the slopes are very steep –
likely unusable for agriculture, but can provide productive conditions for pigs

Geography

Extensive farming is found in the mid-latitude sections of most continents, as well as in desert regions where water for cropping is not available. The nature of extensive farming means it requires less rainfall than intensive farming. The farm is usually large in comparison with the numbers working and money spent on it. In most parts of Western Australia, pastures are so poor that only one sheep to the square mile can be supported

Just as the demand has led to the basic division of cropping and pastoral activities, these areas can also be subdivided depending on the region's rainfall, vegetation type and agricultural activity within the area and the many other parentheses related to this data.

Advantages

1. Extensive farming has a number of advantages over intensive farming:

2. Less labour per unit areas is required to farm large areas, especially since expensive alterations to land (like terracing) are completely absent.

3. Mechanisation can be used more effectively over large, flat areas.

4. Greater efficiency of labour means generally lower product prices.

5. Animal welfare is generally improved because animals are not kept in stifling conditions.

6. Lower requirements of inputs such as fertilizers.

7. If animals are grazed on pastures native to the locality, there is less likely to be problems with exotic species.

8. Local environment and soil are not damaged by overuse of chemicals.

9. The use of machinery and scientific methods of farming produce a large quantity of crops.

Disadvantages

Extensive farming can have the following problems:

1. Yields tend to be much lower than with intensive farming in the short term.

2. Large land requirements limit the habitat of wild species (in some cases, even very low stocking rates can be dangerous), as is the case with intensive farming

Intensive Farming

Intensive farming or intensive agriculture also known as industrial agriculture is characterized by a low fallow ratio and higher use of inputs such as capital and labour per unit land area. This is in contrast to traditional agriculture in which the inputs per unit land are lower.

Intensive animal husbandry involves either large numbers of animals raised on limited land, usually confined animal feeding operations (CAFO) often referred to as factory farms, or managed intensive rotational grazing (MIRG). Both increase the yields of food and fiber per acre as compared to traditional animal husbandry. In a CAFO feed is brought to the animals, which are seldom moved, while in MIRG the animals are repeatedly moved to fresh forage.

Intensive crop agriculture is characterised by innovations designed to increase yield. Techniques include planting multiple crops per year, reducing the frequency of fallow years and improving cultivars. It also involves increased use of fertilizers, plant growth regulators, pesticides and mechanization, controlled by increased and more detailed analysis of growing conditions, including weather, soil, water, weeds and pests.

This system is supported by ongoing innovation in agricultural machinery and farming methods, genetic technology, techniques for achieving economies of scale, logistics and data collection and analysis technology. Intensive farms are widespread in developed nations and increasingly prevalent worldwide. Most of the meat, dairy, eggs, fruits and vegetables available in supermarkets are produced by such farms.

Smaller intensive farms usually include higher inputs of labor and more often use sustainable intensive methods. The farming practices commonly found on such farms are referred to as appropriate technology. These farms are less widespread in both developed countries and worldwide, but are growing more rapidly. Most of the food available in specialty markets such as farmers markets is produced by these smallholder farms.

History

Early 20th-century image of a tractor ploughing an alfalfa field

Agricultural development in Britain between the 16th century and the mid-19th century saw a massive increase in agricultural productivity and net output. This in turn supported unprecedented population growth, freeing up a significant percentage of the workforce, and thereby helped enable the Industrial Revolution. Historians cited enclosure, mechanization, four-field crop rotation, and selective breeding as the most important innovations.

Industrial agriculture arose along with the Industrial Revolution. By the early 19th century, agricultural techniques, implements, seed stocks and cultivars had so improved that yield per land unit was many times that seen in the Middle Ages.

The industrialization phase involved a continuing process of mechanization. Horse-drawn machinery such as the McCormick reaper revolutionized harvesting, while inventions such as the cotton gin reduced the cost of processing. During this same period, farmers began to use steam-powered threshers and tractors, although they were expensive and dangerous. In 1892, the first gasoline-powered tractor was successfully developed, and in 1923, the International Harvester Farmall tractor became the first all-purpose tractor, marking an inflection point in the replacement of draft animals with machines. Mechanical harvesters (combines), planters, transplanters and other equipment were then developed, further revolutionizing agriculture. These inventions increased yields and allowed individual farmers to manage increasingly large farms.

The identification of nitrogen, potassium, and phosphorus (NPK) as critical factors in plant growth led to the manufacture of synthetic fertilizers, further increasing crop yields. In 1909 the Haber-Bosch method to synthesize ammonium nitrate was first demonstrated. NPK fertilizers stimulated the first concerns about industrial agriculture, due to concerns that they came with serious side effects such as soil compaction, soil erosion and declines in overall soil fertility, along with health concerns about toxic chemicals entering the food supply.

The identification of carbon as a critical factor in plant growth and soil health, particularly in the form of humus, led to so-called *sustainable agriculture*, alternative forms of intensive agriculture that also surpass traditional agriculture, without side effects or health issues. Farmers adopting this approach were initially referred to as *humus farmers*, later as *organic farmers*.

The discovery of vitamins and their role in nutrition, in the first two decades of the 20th century, led to vitamin supplements, which in the 1920s allowed some livestock to be raised indoors, reducing their exposure to adverse natural elements. Chemicals developed for use in World War II gave rise to synthetic pesticides.

Following World War II, synthetic fertilizer use increased rapidly, while sustainable intensive farming advanced much more slowly. Most of the resources in developed nations went to improving industrial intensive farming, and very little went to improving organic farming. Thus, particularly in the developed nations, industrial intensive farming grew to become the dominant form of agriculture.

The discovery of antibiotics and vaccines facilitated raising livestock in CAFOs by reducing diseases caused by crowding. Developments in logistics and refrigeration as well as processing technology made long-distance distribution feasible.

Between 1700 and 1980, "the total area of cultivated land worldwide increased 466%" and yields increased dramatically, particularly because of selectively bred high-yielding varieties, fertilizers, pesticides, irrigation and machinery. Global agricultural production doubled between 1820 1920; between 1920 and 1950; between 1950 and 1965; and again between 1965 and 1975 to feed a global population that grew from one billion in 1800 to 6.5 billion in 2002. The number of people involved in farming in industrial countries dropped, from 24 percent of the American population to 1.5 percent in 2002. In 1940, each farmworker supplied 11 consumers, whereas in 2002, each worker supplied 90 consumers. The number of farms also decreased and their ownership became more concentrated. In 2000 in the U.S., four companies produced 81 percent of cows, 73 percent of sheep, 57 percent of pigs, and 50 percent of chickens, cited as an example of "vertical integration" by the president of the U.S. National Farmers' Union. Between 1967 and 2002 the one million pig farms in America consolidated into 114,000 with 80 million pigs (out of 95 million) produced each year on factory farms, according to the U.S. National Pork Producers Council. According to the Worldwatch Institute, 74 percent of the world's poultry, 43 percent of beef, and 68 percent of eggs are produced this way.

Concerns over the sustainability of industrial agriculture, which has become associated with decreased soil quality, and over the environmental effects of fertilizers and pesticides, have not subsided. Alternatives such as integrated pest management (IPM) have had little impact because policies encourage the use of pesticides and IPM is knowledge-intensive. These concerns sustained the organic movement and caused a resurgence in sustainable intensive farming and funding for the development of appropriate technology.

Famines continued throughout the 20th century. Through the effects of climactic events, government policy, war and crop failure, millions of people died in each of at least ten famines between the 1920s and the 1990s.

Techniques and Technologies

Livestock

A commercial chicken house raising broiler pullets for meat.

Confined Animal Feeding Operations

Intensive livestock farming, also called "factory farming" is a term referring to the process of raising livestock in confinement at high stocking density. "Concentrated animal feeding operations" (CAFO) or "intensive livestock operations", can hold large numbers (some up to hundreds of thousands) of cows, hogs, turkeys or chickens, often indoors. The essence of such farms is the concentration of livestock in a given space. The aim is to provide maximum output at the lowest possible cost and with the greatest level of food safety. The term is often used pejoratively. However, CAFOs have dramatically increased the production of food from animal husbandry worldwide, both in terms of total food produced and efficiency.

Food and water is delivered to the animals, and therapeutic use of antimicrobial agents, vitamin supplements and growth hormones are often employed. Growth hormones are not used on chickens nor on any animal in the European Union. Undesirable behaviours often related to the stress of confinement led to a search for docile breeds (e.g., with natural dominance behaviours bred out), physical restraints to stop interaction, such as individual cages for chickens, or physically modification such as the de-beaking of chickens to reduce the harm of fighting.

The CAFO designation resulted from the 1972 US Federal Clean Water Act, which was enacted to protect and restore lakes and rivers to a "fishable, swimmable" quality. The United States Environmental Protection Agency (EPA) identified certain animal feeding operations, along with many other types of industry, as "point source" groundwater polluters. These operations were subjected to regulation.

In 17 states in the U.S., isolated cases of groundwater contamination were linked to CAFOs. For example, the ten million hogs in North Carolina generate 19 million tons of waste per year. The U.S. federal government acknowledges the waste disposal issue and requires that animal waste be stored in lagoons. These lagoons can be as large as 7.5 acres (30,000 m²). Lagoons not protected with an impermeable liner can leak into groundwater under some conditions, as can runoff from manure used as fertilizer. A lagoon that burst in 1995 released 25 million gallons of nitrous sludge in North Carolina's New River. The spill allegedly killed eight to ten million fish.

The large concentration of animals, animal waste and dead animals in a small space poses ethical issues to some consumers. Animal rights and animal welfare activists have charged that intensive animal rearing is cruel to animals.

Other concerns include persistent noxious odor, the effects on human health and the role of antibiotics use in the rise of resistant infectious bacteria.

According to the U.S. Centers for Disease Control and Prevention (CDC), farms on which animals are intensively reared can cause adverse health reactions in farm workers. Workers may develop acute and/or chronic lung disease, musculoskeletal injuries and may catch (zoonotic) infections from the animals.

Managed Intensive Rotational Grazing

Managed Intensive Rotational Grazing (MIRG), also known as cell grazing, mob grazing and holistic managed planned grazing, is a variety of forage use in which herds/flocks are regularly and systematically moved to fresh, rested grazing areas to maximize the quality and quantity of forage growth. MIRG can be used with cattle, sheep, goats, pigs, chickens, turkeys, ducks and other animals. The herds graze one portion of pasture, or a paddock, while allowing the others to recover. Resting grazed lands allows the vegetation to renew energy reserves, rebuild shoot systems, and deepen root systems, resulting in long-term maximum biomass production. MIRG is especially effective because grazers thrive on the more tender younger plant stems. MIRG also leave parasites behind to die off minimizing or eliminating the need for de-wormers. Pasture systems alone can allow grazers to meet their energy requirements, and with the increased productivity of MIRG systems, the animals obtain the majority of their nutritional needs, in some cases all, without the supplemental feed sources that are required in continuous grazing systems or CAFOs.

Crops

The Green Revolution transformed farming in many developing countries. It spread technologies that had already existed, but had not been widely used outside of industrialized nations. These technologies included "miracle seeds", pesticides, irrigation and synthetic nitrogen fertilizer.

Seeds

In the 1970s scientists created strains of maize, wheat, and rice that are generally referred to as high-yielding varieties (HYV). HYVs have an increased nitrogen-absorbing potential compared to other varieties. Since cereals that absorbed extra nitrogen would typically lodge (fall over) before harvest, semi-dwarfing genes were bred into their genomes. Norin 10 wheat, a variety developed by Orville Vogel from Japanese dwarf wheat varieties, was instrumental in developing wheat cultivars. IR8, the first widely implemented HYV rice to be developed by the International Rice Research Institute, was created through a cross between an Indonesian variety named "Peta" and a Chinese variety named "Dee Geo Woo Gen."

With the availability of molecular genetics in Arabidopsis and rice the mutant genes responsible (*reduced height (rht), gibberellin insensitive (gai1)* and *slender rice (slr1)*) have been cloned and identified as cellular signalling components of gibberellic acid, a phytohormone involved in regulating stem growth via its effect on cell division. Photosynthetic investment in the stem is reduced dramatically as the shorter plants are inherently more mechanically stable. Nutrients become redirected to grain production, amplifying in particular the yield effect of chemical fertilisers.

HYVs significantly outperform traditional varieties in the presence of adequate irrigation, pesticides and fertilizers. In the absence of these inputs, traditional varieties may outperform HYVs. They were developed as F1 hybrids, meaning seeds need to be purchased every season to obtain maximum benefit, thus increasing costs.

Crop Rotation

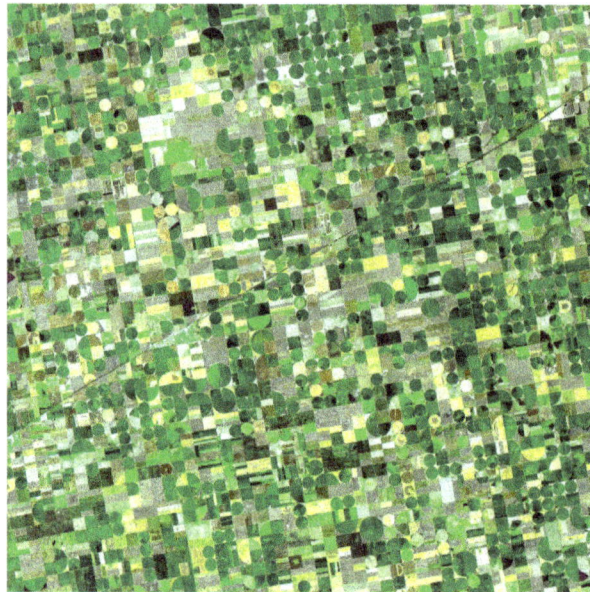

Satellite image of circular crop fields in Haskell County, Kansas in late June 2001. Healthy, growing crops of corn and sorghum are green (Sorghum may be slightly paler). Wheat is brilliant gold. Fields of brown have been recently harvested and plowed under or have lain in fallow for the year.

Crop rotation or crop sequencing is the practice of growing a series of dissimilar types of crops in the same space in sequential seasons for benefits such as avoiding pathogen and pest buildup that occurs when one species is continuously cropped. Crop rotation also seeks to balance the nutrient demands of various crops to avoid soil nutrient depletion. A traditional component of crop rotation is the replenishment of nitrogen through the use of legumes and green manure in sequence with cereals and other crops. Crop rotation can also improve soil structure and fertility by alternating deep-rooted and shallow-rooted plants. One technique is to plant multi-species cover crops between commercial crops. This combines the advantages of intensive farming with continuous cover and polyculture.

Irrigation

Overhead irrigation, center pivot designed

Crop irrigation accounts for 70% of the world's fresh water use.

Flood irrigation, the oldest and most common type, is typically unevenly distributed, as parts of a field may receive excess water in order to deliver sufficient quantities to other parts. Overhead irrigation, using center-pivot or lateral-moving sprinklers, gives a much more equal and controlled distribution pattern. Drip irrigation is the most expensive and least-used type, but delivers water to plant roots with minimal losses.

Water catchment management measures include recharge pits, which capture rainwater and runoff and use it to recharge groundwater supplies. This helps in the replenishment of groundwater wells and eventually reduces soil erosion. Dammed rivers creating Reservoirs store water for irrigation and other uses over large areas. Smaller areas sometimes use irrigation ponds or groundwater.

Weed Control

In agriculture, systematic weed management is usually required, often performed by machines such as cultivators or liquid herbicide sprayers. Herbicides kill specific targets while leaving the crop relatively unharmed. Some of these act by interfering with the growth of the weed and are often based on plant hormones. Weed control through herbicide is made more difficult when the weeds become resistant to the herbicide. Solutions include:

- Cover crops (especially those with allelopathic properties) that out-compete weeds or inhibit their regeneration.

- Multiple herbicides, in combination or in rotation

- Strains genetically engineered for herbicide tolerance

- Locally adapted strains that tolerate or out-compete weeds

- Tilling

- Ground cover such as mulch or plastic

- Manual removal

- Mowing

- Grazing

- Burning

Terracing

Terrace rice fields in Yunnan Province, China

In agriculture, a terrace is a leveled section of a hilly cultivated area, designed as a method of soil conservation to slow or prevent the rapid surface runoff of irrigation water. Often such land is formed into multiple terraces, giving a stepped appearance. The human landscapes of rice cultivation in terraces that follow the natural contours of the escarpments like contour ploughing is a classic feature of the island of Bali and the Banaue Rice Terraces in Banaue, Ifugao, Philippines. In Peru, the Inca made use of otherwise unusable slopes by drystone walling to create terraces.

Rice Paddies

A paddy field is a flooded parcel of arable land used for growing rice and other semiaquatic crops. Paddy fields are a typical feature of rice-growing countries of east and southeast Asia including Malaysia, China, Sri Lanka, Myanmar, Thailand, Korea, Japan, Vietnam, Taiwan, Indonesia, India, and the Philippines. They are also found in other rice-growing regions such as Piedmont (Italy), the Camargue (France) and the Artibonite Valley (Haiti). They can occur naturally along rivers or marshes, or can be constructed, even on hillsides. They require large water quantities for irrigation, much of it from flooding. It gives an environment favourable to the strain of rice being grown, and is hostile to many species of weeds. As the only draft animal species which is comfortable in wetlands, the water buffalo is in widespread use in Asian rice paddies.

Paddy-based rice-farming has been practiced in Korea since ancient times. A pit-house at the Daecheon-ni archaeological site yielded carbonized rice grains and radiocarbon dates indicating that rice cultivation may have begun as early as the Middle Jeulmun Pottery Period (c. 3500-2000 BC) in the Korean Peninsula. The earliest rice cultivation there may have used dry-fields instead of paddies.

The earliest Mumun features were usually located in naturally swampy, low-lying narrow gulleys and fed by local streams. Some Mumun paddies in flat areas were made of a series of squares and rectangles separated by bunds approximately 10 cm in height, while terraced paddies consisted of long irregularly shapes that followed natural contours of the land at various levels.

Like today's, Mumun period rice farmers used terracing, bunds, canals and small reservoirs. Some paddy-farming techniques of the Middle Mumun (c. 850-550 BC) can be interpreted from the well-preserved wooden tools excavated from archaeological rice paddies at the Majeon-ni Site. However, iron tools for paddy-farming were not introduced until sometime after 200 BC. The spatial scale of individual paddies, and thus entire paddy-fields, increased with the regular use of iron tools in the Three Kingdoms of Korea Period (c. AD 300/400-668).

A recent development in the intensive production of rice is System of Rice Intensification (SRI). Developed in 1983 by the French Jesuit Father Henri de Laulanié in Madagascar, by 2013 the number of smallholder farmers using SRI had grown to between 4 and 5 million.

Aquaculture

Aquaculture is the cultivation of the natural products of water (fish, shellfish, algae, seaweed and other aquatic organisms). Intensive aquaculture takes place on land using tanks, ponds or other controlled systems or in the ocean, using cages.

Sustainable Intensive Farming

Sustainable intensive farming practises have been developed to slow the deterioration of agricultural land and even regenerate soil health and ecosystem services, while still offering high yields. Most of these developments fall in the category of organic farming, or the integration of organic and conventional agriculture.

"Organic systems and the practices that make them effective are being picked up more and more by conventional agriculture and will become the foundation for future farming systems. They won't be called organic, because they'll still use some chemicals and still use some fertilizers, but they'll function much more like today's organic systems than today's conventional systems."

Dr. Charles Benbrook Executive director US House Agriculture Subcommittee Director Agricultural Board - National Academy Sciences (FMR).

The System of Crop Intensification (SCI) was born out of research primarily at Cornell University and smallholder farms in India on SRI. It uses the SRI concepts and methods for rice and applies them to crops like wheat, sugarcane, finger millet, and others. It can be 100% organic, or integrated with reduced conventional inputs.

Holistic management is a systems thinking approach that was originally developed for reversing desertification. Holistic planned grazing is similar to rotational grazing but differs in that it more explicitly provides a framework for adapting to four basic ecosystem processes: the water cycle, the mineral cycle including the carbon cycle, energy flow and community dynamics (the relationship between organisms in an ecosystem) as equal in importance to livestock production and social welfare. By intensively managing the behavior and movement of livestock, holistic planned grazing simultaneously increases stocking rates and restores grazing land.

Pasture cropping plants grain crops directly into grassland without first applying herbicides. The perennial grasses form a living mulch understory to the grain crop, eliminating the need to plant cover crops after harvest. The pasture is intensively grazed both before and after grain production using holistic planned grazing. This intensive system yields equivalent farmer profits (partly from increased livestock forage) while building new topsoil and sequestering up to 33 tons of CO_2/ha/year.

The Twelve Aprils grazing program for dairy production, developed in partnership with USDA-SARE, is similar to pasture cropping, but the crops planted into the perennial pasture are forage crops for dairy herds. This system improves milk production and is more sustainable than confinement dairy production.

Integrated Multi-Trophic Aquaculture (IMTA) is an example of a holistic approach. IMTA is a practice in which the by-products (wastes) from one species are recycled to become inputs (fertilizers, food) for another. Fed aquaculture (e.g. fish, shrimp) is combined with inorganic extractive (e.g. seaweed) and organic extractive (e.g. shellfish) aquaculture to create balanced systems for environmental sustainability (biomitigation), economic stability (product diversification and risk reduction) and social acceptability (better management practices).

Biointensive agriculture focuses on maximizing efficiency such as per unit area, energy input and water input. Agroforestry combines agriculture and orchard/forestry technologies to create more integrated, diverse, productive, profitable, healthy and sustainable land-use systems.

Intercropping can increase yields or reduce inputs and thus represents (potentially sustainable) agricultural intensification. However, while total yield per acre is often increased dramatically, yields of any single crop often diminish. There are also challenges to farmers relying on farming equipment optimized for monoculture, often resulting in increased labor inputs.

Vertical farming is intensive crop production on a large scale in urban centers in multi-story, artificially-lit structures that uses far less inputs and produces fewer environmental impacts.

An integrated farming system is a progressive biologically integrated sustainable agriculture system such as IMTA or Zero waste agriculture whose implementation requires exacting knowledge of the interactions of multiple species and whose benefits include sustainability and increased profitability. Elements of this integration can include:

- Intentionally introducing flowering plants into agricultural ecosystems to increase pollen- and nectar-resources required by natural enemies of insect pests
- Using crop rotation and cover crops to suppress nematodes in potatoes

Challenges

The challenges and issues of industrial agriculture for society, for the industrial agriculture sector, for the individual farm, and for animal rights include the costs and benefits of both current practices and proposed changes to those practices. This is a continuation of thousands of years of invention in feeding ever growing populations.

[W]hen hunter-gatherers with growing populations depleted the stocks of game and wild foods across the Near East, they were forced to introduce agriculture. But agriculture brought much longer hours of work and a less rich diet than hunter-gatherers enjoyed. Further population growth among shifting slash-and-burn farmers led to shorter fallow periods, falling yields and soil erosion. Plowing and fertilizers were introduced to deal with these problems - but once again involved longer hours of work and degradation of soil resources(Boserup, The Conditions of Agricultural Growth, Allen and Unwin, 1965, expanded and updated in Population and Technology, Blackwell, 1980.).

While the point of industrial agriculture is to profitably supply the world at the lowest cost, industrial methods have significant side effects. Further, industrial agriculture is not an indivisible whole, but instead is composed of multiple elements, each of which can be modified in response to market conditions, government regulation and further innovation and has its own side-effects. Various interest groups reach different conclusions on the subject.

Benefits

Population Growth

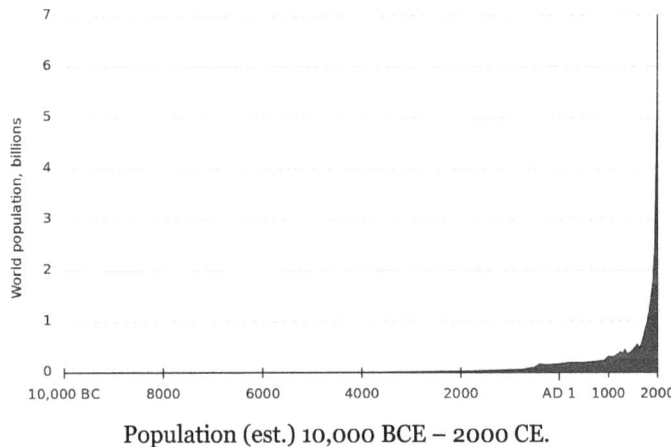

Population (est.) 10,000 BCE – 2000 CE.

Very roughly:

- 30,000 years ago hunter-gatherer behavior fed 6 million people

- 3,000 years ago primitive agriculture fed 60 million people

- 300 years ago intensive agriculture fed 600 million people

- Today industrial agriculture attempts to feed 6 billion people

Estimated world population at various dates, in thousands								
Year	World	Africa	Asia	Europe	Central & South America	North America*	Oceania	Notes
8000 BCE	8 000							
1000 BCE	50 000							

500 BCE	100 000							
1 CE	200,000 plus							
1000	310 000							
1750	791 000	106 000	502 000	163 000	16 000	2 000	2 000	
1800	978 000	107 000	635 000	203 000	24 000	7 000	2 000	
1850	1 262 000	111 000	809 000	276 000	38 000	26 000	2 000	
1900	1 650 000	133 000	947 000	408 000	74 000	82 000	6 000	
1950	2 518 629	221 214	1 398 488	547 403	167 097	171 616	12 812	
1955	2 755 823	246 746	1 541 947	575 184	190 797	186 884	14 265	
1960	2 981 659	277 398	1 674 336	601 401	209 303	204 152	15 888	
1965	3 334 874	313 744	1 899 424	634 026	250 452	219 570	17 657	
1970	3 692 492	357 283	2 143 118	655 855	284 856	231 937	19 443	
1975	4 068 109	408 160	2 397 512	675 542	321 906	243 425	21 564	
1980	4 434 682	469 618	2 632 335	692 431	361 401	256 068	22 828	
1985	4 830 979	541 814	2 887 552	706 009	401 469	269 456	24 678	
1990	5 263 593	622 443	3 167 807	721 582	441 525	283 549	26 687	
1995	5 674 380	707 462	3 430 052	727 405	481 099	299 438	28 924	
2000	6 070 581	795 671	3 679 737	727 986	520 229	315 915	31 043	
2005	6 453 628	887 964	3 917 508	724 722	558 281	332 156	32 998**	

An example of industrial agriculture providing cheap and plentiful food is the U.S.'s "most successful program of agricultural development of any country in the world". Between 1930 and 2000 U.S. agricultural productivity (output divided by all inputs) rose by an average of about 2 percent annually causing food prices to decrease. "The percentage of U.S. disposable income spent on food prepared at home decreased, from 22 percent as late as 1950 to 7 percent by the end of the century."

Liabilities

Environment

Industrial agriculture uses huge amounts of water, energy, and industrial chemicals; increasing pollution in the arable land, usable water and atmosphere. Herbicides, insecticides and fertilizers are accumulating in ground and surface waters. "Many of the negative effects of industrial agriculture are remote from fields and farms. Nitrogen compounds from the Midwest, for example, travel down the Mississippi to degrade coastal fisheries in the Gulf of Mexico. But other adverse effects are showing up within agricultural production systems -- for example, the rapidly developing resistance among pests is rendering our arsenal of herbicides and insecticides increasingly ineffective.". Agrochemicals and monoculture have been implicated in Colony Collapse Disorder, in which the individual members of bee colonies disappear. Agricultural production is highly dependent on bees to pollinate many varieties of fruits and vegetables.

Social

A study done for the US. Office of Technology Assessment conducted by the UC Davis Macrosocial Accounting Project concluded that industrial agriculture is associated with substantial deterioration of human living conditions in nearby rural communities.

Green Revolution

After the Second World War, increased deployment of technologies including pesticides, herbicides, and fertilizers as well as new breeds of high yield crops greatly increased global food production.

The Green Revolution refers to a set of research and development of technology transfer initiatives occurring between the 1930s and the late 1960s (with prequels in the work of the agrarian geneticist Nazareno Strampelli in the 1920s and 1930s), that increased agricultural production worldwide, particularly in the developing world, beginning most markedly in the late 1960s. The initiatives resulted in the adoption of new technologies, including:

...new, high-yielding varieties (HYVs) of cereals, especially dwarf wheats and rices, in association with chemical fertilizers and agro-chemicals, and with controlled water-supply (usually involving irrigation) and new methods of cultivation, including mechanization. All of these together were seen as a 'package of practices' to supersede 'traditional' technology and to be adopted as a whole.

The initiatives, led by Norman Borlaug, the "Father of the Green Revolution," who received the Nobel Peace Prize in 1970, credited with saving over a billion people from starvation, involved the development of high-yielding varieties of cereal grains, expansion of irrigation infrastructure, modernization of management techniques, distribution of hybridized seeds, synthetic fertilizers, and pesticides to farmers.

The term "Green Revolution" was first used in 1968 by former US Agency for International Development (USAID) director William Gaud, who noted the spread of the new technologies: "These and other developments in the field of agriculture contain the makings of a new revolution. It is not a violent Red Revolution like that of the Soviets, nor is it a White Revolution like that of the Shah of Iran. I call it the Green Revolution."

History

Green Revolution in Mexico

It has been argued that "during the twentieth century two 'revolutions' transformed rural Mexico: the Mexican Revolution (1910-1920) and the Green Revolution (1950-1970). With the support of the Mexican government, the U.S. government, the United Nations, the Food and Agriculture Organization (FAO), and the Rockefeller Foundation, Mexico made a concerted effort to transform agricultural productivity, particularly with irrigated rather than dry-land cultivation in its northwest, to solve its problem of lack of food self-sufficiency. In the center and south of Mexico, where large-scale production faced challenges, agricultural production languished. Increased production meant food self-sufficiency in Mexico to feed its growing and urbanizing population, with the number of calories consumed per Mexican increasing.

Mexico was not merely the recipient of Green Revolution knowledge and technology, but was an active participant with financial support from the government for agriculture as well as Mexican agronomists (*agrónomos*). Although the Mexican Revolution had broken the back of the hacienda system and land reform in Mexico had by 1940 distributed a large expanse of land in central and southern Mexico, agricultural productivity had fallen. During the administration of Manuel Avila Camacho (1940–46), the government put resources into developing new breeds of plants and partnered with the Rockefeller Foundation. In 1943, the Mexican government founded the International Maize and Wheat Improvement Center (CIMMYT), which became a base for international agricultural research.

Agriculture in Mexico had been a sociopolitical issue, a key factor in some regions' participation in the Mexican Revolution. It was also a technical issue, which the development of a cohort trained agronomists, who were to advise peasants how to increase productivity. In the post-World War II era, the government sought development in agriculture that bettered technological aspects of agriculture in regions that were not dominated by small-scale peasant cultivators. This drive for transforming agriculture would have the benefit of keeping Mexico self-sufficient in food and in the political sphere with the Cold War, potentially stem unrest and the appeal of Communism. Technical aid can be seen as also serving political ends in the international sphere. In Mexico, it also served political ends, separating peasant agriculture based on the ejido and considered one of the victories of the Mexican Revolution, from agribusiness that requires large-scale land ownership, irrigation, specialized seeds, fertilizers, and pesticides, machinery, and a low-wage paid labor force.

The government created the Mexican Agricultural Program (MAP) to be the lead organization in raising productivity. One of their successes was wheat production, with varieties the agency's scientists helped create dominating wheat production as early as 1951 (70%), 1965 (80%), and 1968 (90%). Mexico became the showcase for extending the Green Revolution to other areas of Latin

America and beyond, into Africa and Asia. New breeds of maize, beans, along with wheat produced bumper crops with proper inputs (such as fertilizer and pesticides) and careful cultivation. Many Mexican farmers who had been dubious about the scientists or hostile to them (often a mutual relationship of discord) came to see the scientific approach to agriculture worth adopting.

Green Revolution in Rice: IR8 and the Philippines

In 1960, the Government of the Republic of the Philippines with the Ford Foundation and the Rockefeller Foundation established IRRI (International Rice Research Institute). A rice crossing between Dee-Geo-woo-gen and Peta was done at IRRI in 1962. In 1966, one of the breeding lines became a new cultivar, IR8. IR8 required the use of fertilizers and pesticides, but produced substantially higher yields than the traditional cultivars. Annual rice production in the Philippines increased from 3.7 to 7.7 million tons in two decades. The switch to IR8 rice made the Philippines a rice exporter for the first time in the 20th century.

Green Revolution's Start in India

In 1961, India was on the brink of mass famine. Norman Borlaug was invited to India by the adviser to the Indian minister of agriculture C. Subramaniam. Despite bureaucratic hurdles imposed by India's grain monopolies, the Ford Foundation and Indian government collaborated to import wheat seed from the International Maize and Wheat Improvement Center (CIMMYT). Punjab was selected by the Indian government to be the first site to try the new crops because of its reliable water supply and a history of agricultural success. India began its own Green Revolution program of plant breeding, irrigation development, and financing of agrochemicals.

India soon adopted IR8 – a semi-dwarf rice variety developed by the International Rice Research Institute (IRRI) that could produce more grains of rice per plant when grown with certain fertilizers and irrigation. In 1968, Indian agronomist S.K. De Datta published his findings that IR8 rice yielded about 5 tons per hectare with no fertilizer, and almost 10 tons per hectare under optimal conditions. This was 10 times the yield of traditional rice. IR8 was a success throughout Asia, and dubbed the "Miracle Rice". IR8 was also developed into Semi-dwarf IR36.

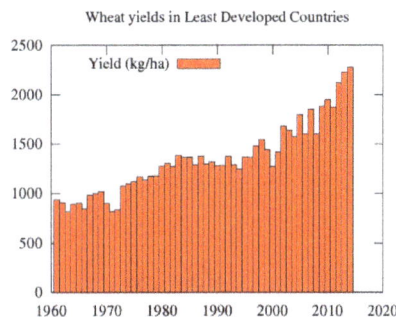

Wheat yields in least developed countries since 1961, in kilograms per hectare.

In the 1960s, rice yields in India were about two tons per hectare; by the mid-1990s, they had risen to six tons per hectare. In the 1970s, rice cost about $550 a ton; in 2001, it cost under $200 a ton. India became one of the world's most successful rice producers, and is now a major rice exporter, shipping nearly 4.5 million tons in 2006.

Consultative Group on International Agricultural Research - CGIAR

In 1970, foundation officials proposed a worldwide network of agricultural research centers under a permanent secretariat. This was further supported and developed by the World Bank; on 19 May 1971, the Consultative Group on International Agricultural Research(CGIAR) was established, co-sponsored by the FAO, IFAD and UNDP. CGIAR, has added many research centers throughout the world.

CGIAR has responded, at least in part, to criticisms of Green Revolution methodologies. This began in the 1980s, and mainly was a result of pressure from donor organizations. Methods like Agroecosystem Analysis and Farming System Research have been adopted to gain a more holistic view of agriculture.

Brazil's Agricultural Revolution

Brazil's vast inland cerrado region was regarded as unfit for farming before the 1960s because the soil was too acidic and poor in nutrients, according to Norman Borlaug. However, from the 1960s, vast quantities of lime (pulverised chalk or limestone) were poured on the soil to reduce acidity. The effort went on and in the late 1990s between 14 million and 16 million tonnes of lime were being spread on Brazilian fields each year. The quantity rose to 25 million tonnes in 2003 and 2004, equalling around five tonnes of lime per hectare. As a result, Brazil has become the world's second biggest soybean exporter and, thanks to the boom in animal feed production, Brazil is now the biggest exporter of beef and poultry in the world.

Problems in Africa

There have been numerous attempts to introduce the successful concepts from the Mexican and Indian projects into Africa. These programs have generally been less successful. Reasons cited include widespread corruption, insecurity, a lack of infrastructure, and a general lack of will on the part of the governments. Yet environmental factors, such as the availability of water for irrigation, the high diversity in slope and soil types in one given area are also reasons why the Green Revolution is not so successful in Africa.

A recent program in western Africa is attempting to introduce a new high-yielding 'family' of rice varieties known as "New Rice for Africa" (NERICA). NERICA varieties yield about 30% more rice under normal conditions, and can double yields with small amounts of fertilizer and very basic irrigation. However, the program has been beset by problems getting the rice into the hands of farmers, and to date the only success has been in Guinea, where it currently accounts for 16% of rice cultivation.

After a famine in 2001 and years of chronic hunger and poverty, in 2005 the small African country of Malawi launched the "Agricultural Input Subsidy Program" by which vouchers are given to smallholder farmers to buy subsidized nitrogen fertilizer and maize seeds. Within its first year, the program was reported to have had extreme success, producing the largest maize harvest of the country's history, enough to feed the country with tons of maize left over. The program has advanced yearly ever since. Various sources claim that the program has been an unusual success, hailing it as a "miracle".

Agricultural Production and Food Security

Technologies

New varieties of wheat and other grains were instrumental to the green revolution.

The Green Revolution spread technologies that already existed, but had not been widely implemented outside industrialized nations. These technologies included modern irrigation projects, pesticides, synthetic nitrogen fertilizer and improved crop varieties developed through the conventional, science-based methods available at the time.

The novel technological development of the Green Revolution was the production of novel wheat cultivars. Agronomists bred cultivars of maize, wheat, and rice that are generally referred to as HYVs or "high-yielding varieties". HYVs have higher nitrogen-absorbing potential than other varieties. Since cereals that absorbed extra nitrogen would typically lodge, or fall over before harvest, semi-dwarfing genes were bred into their genomes. A Japanese dwarf wheat cultivar (Norin 10 wheat), which was sent to Washington, D.C. by Cecil Salmon, was instrumental in developing Green Revolution wheat cultivars. IR8, the first widely implemented HYV rice to be developed by IRRI, was created through a cross between an Indonesian variety named "Peta" and a Chinese variety named "Dee-geo-woo-gen."

With advances in molecular genetics, the mutant genes responsible for *Arabidopsis thaliana* genes (GA 20-oxidase, *ga1*, *ga1-3*), wheat reduced-height genes (*Rht*) and a rice semidwarf gene (*sd1*) were cloned. These were identified as gibberellin biosynthesis genes or cellular signaling component genes. Stem growth in the mutant background is significantly reduced leading to the dwarf phenotype. Photosynthetic investment in the stem is reduced dramatically as the shorter plants are inherently more stable mechanically. Assimilates become redirected to grain production, amplifying in particular the effect of chemical fertilizers on commercial yield.

HYVs significantly outperform traditional varieties in the presence of adequate irrigation, pesticides, and fertilizers. In the absence of these inputs, traditional varieties may outperform HYVs. Therefore, several authors have challenged the apparent superiority of HYVs not only compared to the traditional varieties alone, but by contrasting the monocultural system associated with HYVs with the polycultural system associated with traditional ones.

Production Increases

Cereal production more than doubled in developing nations between the years 1961–1985. Yields of rice, maize, and wheat increased steadily during that period. The production increases can be attributed roughly equally to irrigation, fertilizer, and seed development, at least in the case of Asian rice.

While agricultural output increased as a result of the Green Revolution, the energy input to produce a crop has increased faster, so that the ratio of crops produced to energy input has decreased over time. Green Revolution techniques also heavily rely on chemical fertilizers, pesticides and herbicides and rely on machines, which as of 2014 rely on or are derived from crude oil, making agriculture increasingly reliant on crude oil extraction. Proponents of the Peak Oil theory fear that a future decline in oil and gas production would lead to a decline in food production or even a Malthusian catastrophe.

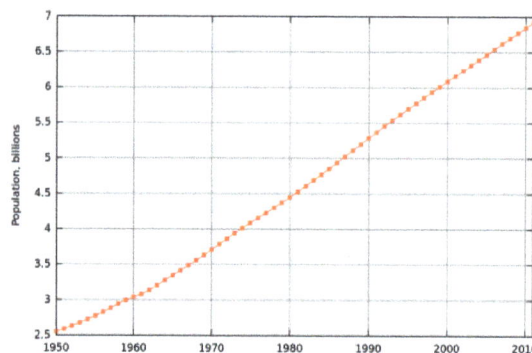

World population 1950–2010

Effects on Food Security

The effects of the Green Revolution on global food security are difficult to assess because of the complexities involved in food systems.

The world population has grown by about four billion since the beginning of the Green Revolution and many believe that, without the Revolution, there would have been greater famine and malnutrition. India saw annual wheat production rise from 10 million tons in the 1960s to 73 million in 2006. The average person in the developing world consumes roughly 25% more calories per day now than before the Green Revolution. Between 1950 and 1984, as the Green Revolution transformed agriculture around the globe, world grain production increased by about 160%.

The production increases fostered by the Green Revolution are often credited with having helped to avoid widespread famine, and for feeding billions of people.

There are also claims that the Green Revolution has decreased food security for a large number of people. One claim involves the shift of subsistence-oriented cropland to cropland oriented towards production of grain for export or animal feed. For example, the Green Revolution replaced much of the land used for pulses that fed Indian peasants for wheat, which did not make up a large portion of the peasant diet.

Criticism

3rd World Economic Sovereignty

A main criticism of the effects of the green revolution is the cost for many small farmers using HYV seeds, with their associated demands of increased irrigation systems and pesticides. A case study is found in India, where farmers are planting cotton seeds capable of producing Bt toxin. A criticism regarding the green revolution are the effects regarding the widespread commercialization and market share of organisations, particularly of the phasing out of seed saving practices in favor of purchasing of seeds, and concerns regarding the financial affordability of the adoption of patented crops amongst farmers, particularly of those in the developing world. This can allow larger farms, even foreign owned farming operations, to buy up local smallhold farms.

Vandana Shiva notes that this is the "second Green Revolution". The first Green Revolution, she notes, was mostly publicly-funded (by the Indian Government). This new Green Revolution, she says, is driven by private [and foreign] interest - notably MNCs like Monsanto. Ultimately, this is leading to foreign ownership over most of India's farmland.

Food Security

Malthusian Criticism

Some criticisms generally involve some variation of the Malthusian principle of population. Such concerns often revolve around the idea that the Green Revolution is unsustainable, and argue that humanity is now in a state of overpopulation or overshoot with regards to the sustainable carrying capacity and ecological demands on the Earth.

Although 36 million people die each year as a direct or indirect result of hunger and poor nutrition, Malthus's more extreme predictions have frequently failed to materialize. In 1798 Thomas Malthus made his prediction of impending famine. The world's population had doubled by 1923 and doubled again by 1973 without fulfilling Malthus's prediction. Malthusian Paul R. Ehrlich, in his 1968 book *The Population Bomb*, said that "India couldn't possibly feed two hundred million more people by 1980" and "Hundreds of millions of people will starve to death in spite of any crash programs." Ehrlich's warnings failed to materialize when India became self-sustaining in cereal production in 1974 (six years later) as a result of the introduction of Norman Borlaug's dwarf wheat varieties.

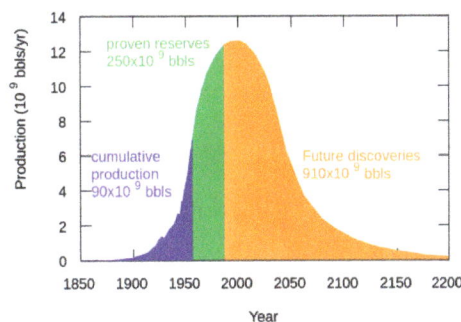

M. King Hubbert's prediction of world petroleum production rates.
Modern agriculture is largely reliant on petroleum energy.

Since supplies of oil and gas are essential to modern agriculture techniques, a fall in global oil supplies could cause spiking food prices in the coming decades.

Famine

To some modern Western sociologists and writers, increasing food production is not synonymous with increasing food security, and is only part of a larger equation. For example, Harvard professor Amartya Sen claimed large historic famines were not caused by decreases in food supply, but by socioeconomic dynamics and a failure of public action. However, economist Peter Bowbrick disputes Sen's theory, arguing that Sen relies on inconsistent arguments and contradicts available information, including sources that Sen himself cited. Bowbrick further argues that Sen's views coincide with that of the Bengal government at the time of the Bengal famine of 1943, and the policies Sen advocates failed to relieve the famine.

Quality of Diet

Some have challenged the value of the increased food production of Green Revolution agriculture. Miguel A. Altieri, (a pioneer of agroecology and peasant-advocate), writes that the comparison between traditional systems of agriculture and Green Revolution agriculture has been unfair, because Green Revolution agriculture produces monocultures of cereal grains, while traditional agriculture usually incorporates polycultures.

These monoculture crops are often used for export, feed for animals, or conversion into biofuel. According to Emile Frison of Bioversity International, the Green Revolution has also led to a change in dietary habits, as fewer people are affected by hunger and die from starvation, but many are affected by malnutrition such as iron or vitamin-A deficiencies. Frison further asserts that almost 60% of yearly deaths of children under age five in developing countries are related to malnutrition.

High-yield rice (HYR), introduced since 1964 to poverty-ridden Asian countries, such as the Philippines, was found to have inferior flavor and be more glutinous and less savory than their native varieties. This caused its price to be lower than the average market value.

In the Philippines the introduction of heavy pesticides to rice production, in the early part of the Green Revolution, poisoned and killed off fish and weedy green vegetables that traditionally coexisted in rice paddies. These were nutritious food sources for many poor Filipino farmers prior to the introduction of pesticides, further impacting the diets of locals.

Political Impact

A major critic of the Green Revolution, U.S. investigative journalist Mark Dowie, writes:

The primary objective of the program was geopolitical: to provide food for the populace in undeveloped countries and so bring social stability and weaken the fomenting of communist insurgency.

Citing internal Foundation documents, Dowie states that the Ford Foundation had a greater concern than Rockefeller in this area.

There is significant evidence that the Green Revolution weakened socialist movements in many nations. In countries such as India, Mexico, and the Philippines, *technological solutions* were sought as an alternative to expanding *agrarian reform* initiatives, the latter of which were often linked to socialist politics.

Socioeconomic Impacts

The transition from traditional agriculture, in which inputs were generated on-farm, to Green Revolution agriculture, which required the purchase of inputs, led to the widespread establishment of rural credit institutions. Smaller farmers often went into debt, which in many cases results in a loss of their farmland. The increased level of mechanization on larger farms made possible by the Green Revolution removed a large source of employment from the rural economy. Because wealthier farmers had better access to credit and land, the Green Revolution increased class disparities, with the rich–poor gap widening as a result. Because some regions were able to adopt Green Revolution agriculture more readily than others (for political or geographical reasons), interregional economic disparities increased as well. Many small farmers are hurt by the dropping prices resulting from increased production overall. However, large-scale farming companies only account for less than 10% of the total farming capacity. This is a criticism held by many small producers in the food sovereignty movement.

The new economic difficulties of small holder farmers and landless farm workers led to increased rural-urban migration. The increase in food production led to a cheaper food for urban dwellers, and the increase in urban population increased the potential for industrialization.

Globalization

In the most basic sense, the Green Revolution was a product of globalization as evidenced in the creation of international agricultural research centers that shared information, and with transnational funding from groups like the Rockefeller Foundation, Ford Foundation, and United States Agency for International Development (USAID).

Environmental Impact

Increased use of irrigation played a major role in the green revolution.

Biodiversity

The spread of Green Revolution agriculture affected both agricultural biodiversity (or agrodiversity) and wild biodiversity. There is little disagreement that the Green Revolution acted to reduce agricultural biodiversity, as it relied on just a few high-yield varieties of each crop.

This has led to concerns about the susceptibility of a food supply to pathogens that cannot be controlled by agrochemicals, as well as the permanent loss of many valuable genetic traits bred into traditional varieties over thousands of years. To address these concerns, massive seed banks such as Consultative Group on International Agricultural Research's (CGIAR) International Plant Genetic Resources Institute (now Bioversity International) have been established.

There are varying opinions about the effect of the Green Revolution on wild biodiversity. One hypothesis speculates that by increasing production per unit of land area, agriculture will not need to expand into new, uncultivated areas to feed a growing human population. However, land degradation and soil nutrients depletion have forced farmers to clear up formerly forested areas in order to keep up with production. A counter-hypothesis speculates that biodiversity was sacrificed because traditional systems of agriculture that were displaced sometimes incorporated practices to preserve wild biodiversity, and because the Green Revolution expanded agricultural development into new areas where it was once unprofitable or too arid. For example, the development of wheat varieties tolerant to acid soil conditions with high aluminium content, permitted the introduction of agriculture in sensitive Brazilian ecosystems as Cerrado semi-humid tropical savanna and Amazon rainforest in the geoeconomic macroregions of Centro-Sul and Amazônia. Before the Green Revolution, other Brazilian ecosystems were also significantly damaged by human activity, such as the once 1st or 2nd main contributor to Brazilian megadiversity Atlantic Rainforest (above 85% of deforestation in the 1980s, about 95% after the 2010s) and the important xeric shrublands called Caatinga mainly in the Northeastern Brazil (about 40% in the 1980s, about 50% after the 2010s — deforestation of the Caatinga biome is generally associated with greater risks of desertification).

Nevertheless, the world community has clearly acknowledged the negative aspects of agricultural expansion as the 1992 Rio Treaty, signed by 189 nations, has generated numerous national Biodiversity Action Plans which assign significant biodiversity loss to agriculture's expansion into new domains.

Greenhouse Gas Emissions

According to a study published in 2013 in PNAS, in the absence of the crop germplasm improvement associated with the Green revolution, greenhouse gas emissions would have been 5.2-7.4 Gt higher than observed in 1965–2004.

Dependence on Non-renewable Resources

Most high intensity agricultural production is highly reliant on non-renewable resources. Agricultural machinery and transport, as well as the production of pesticides and nitrates all depend on fossil fuels. Moreover, the essential mineral nutrient phosphorus is often a limiting factor in crop cultivation, while phosphorus mines are rapidly being depleted worldwide. The failure to depart from these non-sustainable agricultural production methods could potentially lead to a large scale collapse of the current system of intensive food production within this century.

Health Impact

The consumption of the pesticides used to kill pests by humans in some cases may be increasing the likelihood of cancer in some of the rural villages using them. Poor farming practices including non-compliance to usage of masks and over-usage of the chemicals compound this situation. In 1989, WHO and UNEP estimated that there were around 1 million human pesticide poisonings annually. Some 20,000 (mostly in developing countries) ended in death, as a result of poor labeling, loose safety standards etc.

Pesticides and Cancer

Long term exposure to pesticides such as organochlorines, creosote, and sulfate have been correlated with higher cancer rates and organochlorines DDT, chlordane, and lindane as tumor promoters in animals. Contradictory epidemiologic studies in humans have linked phenoxy acid herbicides or contaminants in them with soft tissue sarcoma (STS) and malignant lymphoma, organochlorine insecticides with STS, non-Hodgkin's lymphoma (NHL), leukemia, and, less consistently, with cancers of the lung and breast, organophosphorous compounds with NHL and leukemia, and triazine herbicides with ovarian cancer.

Punjab Case

The Indian state of Punjab pioneered green revolution among the other states transforming India into a food-surplus country. The state is witnessing serious consequences of intensive farming using chemicals and pesticide. A comprehensive study conducted by Post Graduate Institute of Medical Education and Research (PGIMER) has underlined the direct relationship between indiscriminate use of these chemicals and increased incidence of cancer in this region. An increase in the number of cancer cases has been reported in several villages including Jhariwala, Koharwala, Puckka, Bhimawali, and Khara.

Environmental activist Vandana Shiva has written extensively about the social, political and economic impacts of the Green Revolution in Punjab. She claims that the Green Revolution's reliance on heavy use of chemical inputs and monocultures has resulted in water scarcity, vulnerability to pests, and incidents of violent conflict and social marginalization.

In 2009, under a Greenpeace Research Laboratories investigation, Dr Reyes Tirado, from the University of Exeter, UK conducted the study in 50 villages in Muktsar, Bathinda and Ludhiana districts revealed chemical, radiation and biological toxicity rampant in Punjab. Twenty percent of the sampled wells showed nitrate levels above the safety limit of 50 mg/l, established by WHO, the study connected it with high use of synthetic nitrogen fertilizers.

Norman Borlaug's Response to Criticism

Borlaug dismissed certain claims of critics, but also cautioned "There are no miracles in agricultural production. Nor is there such a thing as a miracle variety of wheat, rice, or maize which can serve as an elixir to cure all ills of a stagnant, traditional agriculture."

Of environmental lobbyists, he said:

"some of the environmental lobbyists of the Western nations are the salt of the earth, but many of them are elitists. They've never experienced the physical sensation of hunger. They do their lobbying from comfortable office suites in Washington or Brussels...If they lived just one month amid the misery of the developing world, as I have for fifty years, they'd be crying out for tractors and fertilizer and irrigation canals and be outraged that fashionable elitists back home were trying to deny them these things".

However, the charge of "elitism" could also be leveled against supporters of the Green Revolution. As noted in the documentary Profits from Poison, protective gear for farmers is often designed by people who have never experienced tropical climates. For example, plastic ponchos create a sauna effect if worn in high temperature/humidity, unlike the experience of wearing them in an airconditioned office.

The "New" Green Revolution

Although the Green Revolution has been able to improve agricultural output in some regions in the world, there was and is still room for improvement. As a result, many organizations continue to invent new ways to improve the techniques already used in the Green Revolution. Frequently quoted inventions are the System of Rice Intensification, marker-assisted selection, agroecology, and applying existing technologies to agricultural problems of the developing world.

References

- F. Stuart Chapin III; Pamela A. Matson; Harold A. Moon (2002). Principles of Terrestrial Ecosystem Ecology. Springer. ISBN 0387954392.

- Kötke, William H. (1993). The Final Empire: The Collapse of Civilization and the Seed of the Future. Arrow Point Press. ISBN 0963378457.

- Stinner, D.H (2007). "The Science of Organic Farming". In William Lockeretz. Organic Farming: An International History. Oxfordshire, UK & Cambridge, Massachusetts: CAB International (CABI). ISBN 978-0-85199-833-6. Retrieved 30 April 2013 ebook ISBN 978-1-84593-289-3

- American Foundations: An Investigative History; Dowie, Mark; 13 April 2001; MIT Press; Massachusetts; (retrieved from Goodreads online); ISBN 0262041898; accessed March 2014.

- "The System of Crop Intensification Agroecological Innovations for Improving Agricultural Production, Food Security, and Resilience to Climate Change" (PDF). SRI International Network and Resources Center. Cornell University. Retrieved 1 October 2014.

- Schwartz, Judith D. "Soil as Carbon Storehouse: New Weapon in Climate Fight?". Yale Environment 360. Yale School of Forestry & Environmental Studies. Retrieved 25 June 2014.

- Leu, Andre. "Mitigating Climate Change With Soil Organic Matter in Organic Production Systems" (PDF). Trade and environment review 2013, Commentary V pp.22-32. UNCTAD. Retrieved 28 September 2014.

- "Green Revolution research saved an estimated 18 to 27 million hectares from being brought into agricultural production". Pnas.org. 2013-05-13. Retrieved 2013-08-28.

- Philpott, Tom (19 April 2013). "A Brief History of Our Deadly Addiction to Nitrogen Fertilizer". Mother Jones. Retrieved 2013-05-07.

- Undersander, Dan; et al. "Pastures for profit: A guide to rotational grazing" (PDF). University of Wisconsin Extension. Retrieved 5 April 2013.

- link)Fairlie, Simon. "Maximizing Soil Carbon Sequestration: Carbon Farming and Rotational Grazing". Mother Earth News August 21, 2012. Retrieved 7 April 2013.

- "FIRST MILLIMETER: HEALING THE EARTH PREVIOUS BROADCASTS". KQED PUBLIC MEDIA FOR NORTHERN CALIFORNIA. Retrieved 20 April 2013.

- Gaud, William S. (8 March 1968). "The Green Revolution<3 Accomplishments and Apprehensions". AgBio-World. Retrieved 8 August 2011.

Pesticides, Herbicides and Weed Control

Weeding and pest management are a very important part of agronomic practice. Agricultural pests do great harm to agricultural produce. The chapter touches upon pesticides, herbicides and weed control. The chapter discusses the methods of agronomy in a critical manner providing key analysis to the subject matter.

Weed Control

Weed control is the botanical component of pest control, which attempts to stop weeds, especially noxious or injurious weeds, from competing with domesticated plants and livestock. Many strategies have been developed in order to contain these plants.

The original strategy was manual removal including ploughing, which can cut the roots of weeds. More recent approaches include herbicides (chemical weed killers) and reducing stocks by burning and/or pulverizing seeds.

A plant is often termed a "weed" when it has one or more of the following characteristics:

- Little or no recognized value (as in medicinal, material, nutritional or energy)
- Rapid growth and/or ease of germination
- Competitive with crops for space, light, water and nutrients

The definition of a weed is completely context-dependent. To one person, one plant may be a weed, and to another person it may be a desirable plant. In one place, a plant may be viewed as a weed, whereas in another place, the same plant may be desirable.

Introduction

Weeds compete with productive crops or pasture, ultimately converting productive land into unusable scrub. Weeds can be poisonous, distasteful, produce burrs, thorns or otherwise interfere with the use and management of desirable plants by contaminating harvests or interfering with livestock.

Weeds compete with crops for space, nutrients, water and light. Smaller, slower growing seedlings are more susceptible than those that are larger and more vigorous. Onions are one of the most vulnerable, because they are slow to germinate and produce slender, upright stems. By contrast broad beans produce large seedlings and suffer far fewer effects other than during periods of water shortage at the crucial time when the pods are filling out. Transplanted crops raised in sterile soil or potting compost gain a head start over germinating weeds.

Weeds also vary in their competitive abilities and according to conditions and season. Tall-growing vigorous weeds such as fat hen (*Chenopodium album*) can have the most pronounced effects on adjacent crops, although seedlings of fat hen that appear in late summer produce only small plants. Chickweed (*Stellaria media*), a low growing plant, can happily co-exist with a tall crop during the summer, but plants that have overwintered will grow rapidly in early spring and may swamp crops such as onions or spring greens.

The presence of weeds does not necessarily mean that they are damaging a crop, especially during the early growth stages when both weeds and crops can grow without interference. However, as growth proceeds they each begin to require greater amounts of water and nutrients. Estimates suggest that weed and crop can co-exist harmoniously for around three weeks before competition becomes significant. One study found that after competition had started, the final yield of onion bulbs was reduced at almost 4% per day.

Perennial weeds with bulbils, such as lesser celandine and oxalis, or with persistent underground stems such as couch grass (*Agropyron repens*) or creeping buttercup (*Ranunculus repens*) store reserves of food, and are thus able to grow faster and with more vigour than their annual counterparts. Some perennials such as couch grass exude allelopathic chemicals that inhibit the growth of other nearby plants.

Weeds can also host pests and diseases that can spread to cultivated crops. Charlock and Shepherd's purse may carry clubroot, eelworm can be harboured by chickweed, fat hen and shepherd's purse, while the cucumber mosaic virus, which can devastate the cucurbit family, is carried by a range of different weeds including chickweed and groundsel.

Insect pests often do not attack weeds. However pests such as cutworms may first attack weeds then move on to cultivated crops.

Some plants are considered weeds by some farmers and crops by others. Charlock, a common weed in the southeastern US, are weeds according to row crop growers, but are valued by beekeepers, who seek out places where it blooms all winter, thus providing pollen for honeybees and other pollinators. Its bloom resists all but a very hard freeze, and recovers once the freeze ends.

Weed Propagation

Seeds

Annual and biennial weeds such as chickweed, annual meadow grass, shepherd's purse, groundsel, fat hen, cleaver, speedwell and hairy bittercress propagate themselves by seeding. Many produce huge numbers of seed several times a season, some all year round. Groundsel can produce 1000 seed, and can continue right through a mild winter, whilst Scentless Mayweedproduces over 30,000 seeds per plant. Not all of these will germinate at once, but over several seasons, lying dormant in the soil sometimes for years until exposed to light. Poppy seed can survive 80–100 years, dock 50 or more. There can be many thousands of seeds in a square foot or square metre of ground, thus and soil disturbance will produce a flush of fresh weed seedlings.

Subsurface/Surface

The most persistent perennials spread by underground creeping rhizomes that can regrow from a tiny fragment. These include couch grass, bindweed, ground elder, nettles, rosebay willow herb, Japanese knotweed, horsetail and bracken, as well as creeping thistle, whose tap roots can put out lateral roots. Other perennials put out runners that spread along the soil surface. As they creep they set down roots, enabling them to colonise bare ground with great rapidity. These include creeping buttercup and ground ivy. Yet another group of perennials propagate by stolons- stems that arch back into the ground to reroot. The most familiar of these is the bramble.

Methods

Weed control plans typically consist of many methods which are divided into biological, chemical, cultural, and physical/mechanical control.

Pesticide-free thermic weed control with a weed burner on a potato field in Dithmarschen

Physical/Mechanical Methods

Coverings

In domestic gardens, methods of weed control include covering an area of ground with a material that creates a hostile environment for weed growth, known as a *weed mat*.

Several layers of wet newspaper prevent light from reaching plants beneath, which kills them. Daily saturating the newspaper with water plant decomposition. After several weeks, all germinating weed seeds are dead.

In the case of black plastic, the greenhouse effect kills the plants. Although the black plastic sheet is effective at preventing weeds that it covers, it is difficult to achieve complete coverage. Eradicating persistent perennials may require the sheets to be left in place for at least two seasons.

Some plants are said to produce root exudates that suppress herbaceous weeds. *Tagetes minuta* is claimed to be effective against couch and ground elder, whilst a border of comfrey is also said to act as a barrier against the invasion of some weeds including couch. A 5–10 centimetres (2.0–3.9 in)} layer of wood chip mulch prevents most weeds from sprouting.

Gravel can serve as an inorganic mulch.

Irrigation is sometimes used as a weed control measure such as in the case of paddy fields to kill any plant other than the water-tolerant rice crop.

Manual Removal

Weeds are removed manually in large parts of India.

Many gardeners still remove weeds by manually pulling them out of the ground, making sure to include the roots that would otherwise allow them to resprout.

Hoeing off weed leaves and stems as soon as they appear can eventually weaken and kill perennials, although this will require persistence in the case of plants such as bindweed. Nettle infestations can be tackled by cutting back at least three times a year, repeated over a three-year period. Bramble can be dealt with in a similar way.

Tillage

Ploughing includes tilling of soil, intercultural ploughing and summer ploughing. Ploughing uproots weeds, causing them to die. In summer ploughing is done during deep summers. Summer ploughing also helps in killing pests.

Mechanical tilling can remove weeds around crop plants at various points in the growing process.

Thermal

Several thermal methods can control weeds.

Hot foam (foamstream) causes the cell walls to rupture, killing the plant. Weed burners heat up soil quickly and destroy superficial parts of the plants. Weed seeds are often heat resistant and even react with an increase of growth on dry heat.

Since the 19th century soil steam sterilization has been used to clean weeds completely from soil. Several research results confirm the high effectiveness of humid heat against weeds and its seeds.

Soil solarization in some circumstances is very effective at eliminating weeds while maintaining grass. Planted grass tends to have a higher heat/humidity tolerance than unwanted weeds.

Seed Targeting

In 1998, the Australian Herbicide Resistance Initiative (AHRI), debuted. gathered fifteen scientists and technical staff members to conduct field surveys, collect seeds, test for resistance and study the biochemical and genetic mechanisms of resistance. A collaboration with DuPont led to a mandatory herbicide labeling program, in which each mode of action is clearly identified by a letter of the alphabet.

The key innovation of the AHRI approach has been to focus on weed seeds. Ryegrass seeds last only a few years in soil, so if farmers can prevent new seeds from arriving, the number of sprouts will shrink each year. Until the new approach farmers were unintentionally helping the seeds. Their combines loosen ryegrass seeds from their stalks and spread them over the fields. In the mid-1980s, a few farmers hitched covered trailers, called "chaff carts", behind their combines to catch the chaff and weed seeds. The collected material is then burned.

An alternative is to concentrate the seeds into a half-meter-wide strip called a windrow and burn the windrows after the harvest, destroying the seeds. Since 2003, windrow burning has been adopted by about 70% of farmers in Western Australia.

Yet another approach is the Harrington Seed Destructor, which is an adaptation of a coal pulverizing cage mill that uses steel bars whirling at up to 1500 rpm. It keeps all the organic material in the field and does not involve combustion, but kills 95% of seeds.

Cultural Methods

Stale Seed Bed

Another manual technique is the 'stale seed bed', which involves cultivating the soil, then leaving it fallow for a week or so. When the initial weeds sprout, the grower lightly hoes them away before planting the desired crop. However, even a freshly cleared bed is susceptible to airborne seed from elsewhere, as well as seed carried by passing animals on their fur, or from imported manure.

Buried Drip Irrigation

Buried drip irrigation involves burying drip tape in the subsurface near the planting bed, thereby limiting weeds access to water while also allowing crops to obtain moisture. It is most effective during dry periods.

Crop Rotation

Rotating crops with ones that kill weeds by choking them out, such as hemp, *Mucuna pruriens*, and other crops, can be a very effective method of weed control. It is a way to avoid the use of herbicides, and to gain the benefits of crop rotation.

Biological Methods

A biological weed control regiment can consist of biological control agents, bioherbicides, use of grazing animals, and protection of natural predators.

Animal Grazing

Companies using goats to control and eradicate leafy spurge, knapweed, and other toxic weeds have sprouted across the American West.

Chemical Methods

"Organic" Approaches

Weed control, circa 1930-40s

A mechanical weed control device: the diagonal weeder

Organic weed control involves anything other than applying manufactured chemicals. Typically a combination of methods are used to achieve satisfactory control.

Sulfur in some circumstances is accepted within British Soil Association standards.

Herbicides

The above described methods of weed control use no or very limited chemical inputs. They are preferred by organic gardeners or organic farmers.

However weed control can also be achieved by the use of herbicides. Selective herbicides kill certain targets while leaving the desired crop relatively unharmed. Some of these act by interfering with the growth of the weed and are often based on plant hormones. Herbicides are generally classified as follows:

- Contact herbicides destroy only plant tissue that contacts the herbicide. Generally, these are the fastest-acting herbicides. They are ineffective on perennial plants that can re-grow from roots or tubers.

- Systemic herbicides are foliar-applied and move through the plant where they destroy a greater amount of tissue. Glyphosate is currently the most used systemic herbicide.

- Soil-borne herbicides are applied to the soil and are taken up by the roots of the target plant.

- Pre-emergent herbicides are applied to the soil and prevent germination or early growth of weed seeds.

In agriculture large scale and systematic procedures are usually required, often by machines, such as large liquid herbicide 'floater' sprayers, or aerial application.

Bradley Method

Bradley Method of Bush Regeneration uses ecological processes to do much of the work. Perennial weeds also propagate by seeding; the airborne seed of the dandelion and the rose-bay willow herb parachute far and wide. Dandelion and dock also put down deep tap roots, which, although they do not spread underground, are able to regrow from any remaining piece left in the ground.

Hybrid

One method of maintaining the effectiveness of individual strategies is to combine them with others that work in complete different ways. Thus seed targeting has been combined with herbicides. In Australia seed management has been effectively combined with trifluralin and clethodim.

Resistance

Resistance occurs when a target adapts to circumvent a particular control strategy. It affects not only weed control,but antibiotics, insect control and other domains. In agriculture is mostly considered in reference to pesticides, but can defeat other strategies, e.g., when a target species becomes more drought tolerant via selection pressure.

Farming Practices

Herbicide resistance recently became a critical problem as many Australian sheep farmers switched to exclusively growing wheat in their pastures in the 1970s. In wheat fields, introduced varieties of ryegrass, while good for grazing sheep, are intense competitors with wheat. Ryegrasses produce so many seeds that, if left unchecked, they can completely choke a field. Herbicides provided excellent control, while reducing soil disrupting because of less need to plough. Within little more than a decade, ryegrass and other weeds began to develop resistance. Australian farmers evolved again and began diversifying their techniques.

In 1983, patches of ryegrass had become immune to Hoegrass, a family of herbicides that inhibit an enzyme called acetyl coenzyme A carboxylase.

Ryegrass populations were large, and had substantial genetic diversity, because farmers had planted many varieties. Ryegrass is cross-pollinated by wind, so genes shuffle frequently. Farmers sprayed inexpensive Hoegrass year after year, creating selection pressure, but were diluting the herbicide in order to save money, increasing plants survival. Hoegrass was mostly replaced by a group of herbicides that block acetolactate synthase, again helped by poor application practices. Ryegrass evolved a kind of "cross-resistance" that allowed it to rapidly break down a variety of herbicides. Australian farmers lost four classes of herbicides in only a few years. As of 2013 only two herbicide classes, called Photosystem II and long-chain fatty acid inhibitors, had become the last hope.

Legislation

A crop-duster spraying pesticide on a field

A Lite-Trac four-wheeled self-propelled crop sprayer spraying pesticide on a field

Pesticides are substances meant for attracting, seducing, and then destroying any pest. They are a class of biocide. The most common use of pesticides is as plant protection products (also known as crop protection products), which in general protect plants from damaging influences such as weeds, fungi, or insects. This use of pesticides is so common that the term *pesticide* is often treated as synonymous with *plant protection product*, although it is in fact a broader term, as pesticides are also used for non-agricultural purposes. The term pesticide includes all of the following: herbicide, insecticide, insect growth regulator, nematicide, termiticide, molluscicide, piscicide, avicide, rodenticide, predacide, bactericide, insect repellent, animal repellent, antimicrobial, fungicide, disinfectant (antimicrobial), and sanitizer.

In general, a pesticide is a chemical or biological agent (such as a virus, bacterium, antimicrobial, or disinfectant) that deters, incapacitates, kills, or otherwise discourages pests. Target pests can include insects, plant pathogens, weeds, mollusks, birds, mammals, fish, nematodes (roundworms), and microbes that destroy property, cause nuisance, or spread disease, or are disease vectors. Although pesticides have benefits, some also have drawbacks, such as potential toxicity to humans and other species. According to the Stockholm Convention on Persistent Organic Pollutants, 9 of the 12 most dangerous and persistent organic chemicals are organochlorine pesticides.

Definition

Type of pesticide	Target pest group
Herbicides	Plant
Algicides or Algaecides	Algae
Avicides	Birds
Bactericides	Bacteria
Fungicides	Fungi and Oomycetes
Insecticides	Insects
Miticides or Acaricides	Mites
Molluscicides	Snails
Nematicides	Nematodes
Rodenticides	Rodents
Virucides	Viruses

The Food and Agriculture Organization (FAO) has defined *pesticide* as:

> any substance or mixture of substances intended for preventing, destroying, or controlling any pest, including vectors of human or animal disease, unwanted species of plants or animals, causing harm during or otherwise interfering with the production, processing, storage, transport, or marketing of food, agricultural commodities, wood and wood products or animal feedstuffs, or substances that may be administered to animals for the control of insects, arachnids, or other pests in or on their bodies. The term includes substances intended for use as a plant growth regulator, defoliant, desiccant, or agent for thinning fruit or preventing the premature fall of fruit. Also used as substances applied to crops either before or after harvest to protect the commodity from deterioration during storage and transport.

Pesticides can be classified by target organism (e.g., herbicides, insecticides, fungicides, rodenticides, and pediculicides), chemical structure (e.g., organic, inorganic, synthetic, or biological (biopesticide), although the distinction can sometimes blur), and physical state (e.g. gaseous (fumigant)). Biopesticides include microbial pesticides and biochemical pesticides. Plant-derived pesticides, or "botanicals", have been developing quickly. These include the pyrethroids, rotenoids, nicotinoids, and a fourth group that includes strychnine and scilliroside.

Many pesticides can be grouped into chemical families. Prominent insecticide families include organochlorines, organophosphates, and carbamates. Organochlorine hydrocarbons (e.g., DDT) could be separated into dichlorodiphenylethanes, cyclodiene compounds, and other related compounds. They operate by disrupting the sodium/potassium balance of the nerve fiber, forcing the nerve to transmit continuously. Their toxicities vary greatly, but they have been phased out because of their persistence and potential to bioaccumulate. Organophosphate and carbamates largely replaced organochlorines. Both operate through inhibiting the enzyme acetylcholinesterase, allowing acetylcholine to transfer nerve impulses indefinitely and causing a variety of symptoms such as weakness or paralysis. Organophosphates are quite toxic to vertebrates, and have in some cases been replaced by less toxic carbamates. Thiocarbamate and dithiocarbamates are subclasses of carbamates. Prominent families of herbicides include phenoxy and benzoic acid herbicides (e.g. 2,4-D), triazines (e.g., atrazine), ureas (e.g., diuron), and Chloroacetanilides (e.g., alachlor). Phenoxy compounds tend to selectively kill broad-leaf weeds rather than grasses. The phenoxy and benzoic acid herbicides function similar to plant growth hormones, and grow cells without normal cell division, crushing the plant's nutrient transport system. Triazines interfere with photosynthesis. Many commonly used pesticides are not included in these families, including glyphosate.

Pesticides can be classified based upon their biological mechanism function or application method. Most pesticides work by poisoning pests. A systemic pesticide moves inside a plant following absorption by the plant. With insecticides and most fungicides, this movement is usually upward (through the xylem) and outward. Increased efficiency may be a result. Systemic insecticides, which poison pollen and nectar in the flowers, may kill bees and other needed pollinators.

In 2009, the development of a new class of fungicides called paldoxins was announced. These work by taking advantage of natural defense chemicals released by plants called phytoalexins, which fungi then detoxify using enzymes. The paldoxins inhibit the fungi's detoxification enzymes. They are believed to be safer and greener.

Uses

Pesticides are used to control organisms that are considered to be harmful. For example, they are used to kill mosquitoes that can transmit potentially deadly diseases like West Nile virus, yellow fever, and malaria. They can also kill bees, wasps or ants that can cause allergic reactions. Insecticides can protect animals from illnesses that can be caused by parasites such as fleas. Pesticides can prevent sickness in humans that could be caused by moldy food or diseased produce. Herbicides can be used to clear roadside weeds, trees and brush. They can also kill invasive weeds that may cause environmental damage. Herbicides are commonly applied in ponds and lakes to control algae and plants such as water grasses that can interfere with activities like swimming and fishing and cause the water to look or smell unpleasant. Uncontrolled pests such as termites and mold can damage structures such as houses. Pesticides are used in grocery stores and food storage facilities to manage rodents and insects that infest food such as grain. Each use of a pesticide carries some associated risk. Proper pesticide use decreases these associated risks to a level deemed acceptable by pesticide regulatory agencies such as the United States Environmental Protection Agency (EPA) and the Pest Management Regulatory Agency (PMRA) of Canada.

DDT, sprayed on the walls of houses, is an organochlorine that has been used to fight malaria since the 1950s. Recent policy statements by the World Health Organization have given stronger support to this approach. However, DDT and other organochlorine pesticides have been banned in most countries worldwide because of their persistence in the environment and human toxicity. DDT use is not always effective, as resistance to DDT was identified in Africa as early as 1955, and by 1972 nineteen species of mosquito worldwide were resistant to DDT.

Amount Used

In 2006 and 2007, the world used approximately 2.4 megatonnes (5.3×10^9 lb) of pesticides, with herbicides constituting the biggest part of the world pesticide use at 40%, followed by insecticides (17%) and fungicides (10%). In 2006 and 2007 the U.S. used approximately 0.5 megatonnes (1.1×10^9 lb) of pesticides, accounting for 22% of the world total, including 857 million pounds (389 kt) of conventional pesticides, which are used in the agricultural sector (80% of conventional pesticide use) as well as the industrial, commercial, governmental and home & garden sectors. Pesticides are also found in majority of U.S. households with 78 million out of the 105.5 million households indicating that they use some form of pesticide. As of 2007, there were more than 1,055 active ingredients registered as pesticides, which yield over 20,000 pesticide products that are marketed in the United States.

The US used some 1 kg (2.2 pounds) per hectare of arable land compared with: 4.7 kg in China, 1.3 kg in the UK, 0.1 kg in Cameroon, 5.9 kg in Japan and 2.5 kg in Italy. Insecticide use in the US has declined by more than half since 1980, (.6%/yr) mostly due to the near phase-out of organophosphates. In corn fields, the decline was even steeper, due to the switchover to transgenic Bt corn.

For the global market of crop protection products, market analysts forecast revenues of over 52 billion US$ in 2019.

Benefits

Pesticides can save farmers' money by preventing crop losses to insects and other pests; in the U.S., farmers get an estimated fourfold return on money they spend on pesticides. One study found that not using pesticides reduced crop yields by about 10%. Another study, conducted in 1999, found that a ban on pesticides in the United States may result in a rise of food prices, loss of jobs, and an increase in world hunger.

There are two levels of benefits for pesticide use, primary and secondary. Primary benefits are direct gains from the use of pesticides and secondary benefits are effects that are more long-term.

Primary Benefits

1. Controlling pests and plant disease vectors

    - Improved crop/livestock yields

    - Improved crop/livestock quality

    - Invasive species controlled

2. Controlling human/livestock disease vectors and nuisance organisms

    - Human lives saved and suffering reduced

    - Animal lives saved and suffering reduced

    - Diseases contained geographically

3. Controlling organisms that harm other human activities and structures

    - Drivers view unobstructed

    - Tree/brush/leaf hazards prevented

    - Wooden structures protected

Monetary

Every dollar ($1) that is spent on pesticides for crops yields four dollars ($4) in crops saved. This means based that, on the amount of money spent per year on pesticides, $10 billion, there is an additional $40 billion savings in crop that would be lost due to damage by insects and weeds. In general, farmers benefit from having an increase in crop yield and from being able to grow a variety of crops throughout the year. Consumers of agricultural products also benefit from being able to afford the vast quantities of produce available year-round. The general public also benefits from the use of pesticides for the control of insect-borne diseases and illnesses, such as malaria. The use of pesticides creates a large job market within the agrichemical sector.

Costs

On the cost side of pesticide use there can be costs to the environment, costs to human health, as well as costs of the development and research of new pesticides.

Health Effects

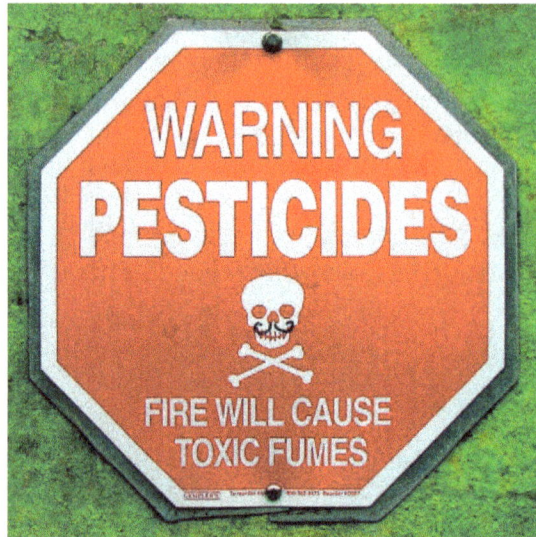

A sign warning about potential pesticide exposure.

Pesticides may cause acute and delayed health effects in people who are exposed. Pesticide exposure can cause a variety of adverse health effects, ranging from simple irritation of the skin and eyes to more severe effects such as affecting the nervous system, mimicking hormones causing reproductive problems, and also causing cancer. A 2007 systematic review found that "most studies on non-Hodgkin lymphoma and leukemia showed positive associations with pesticide exposure" and thus concluded that cosmetic use of pesticides should be decreased. There is substantial evidence of associations between organophosphate insecticide exposures and neurobehavioral alterations. Limited evidence also exists for other negative outcomes from pesticide exposure including neurological, birth defects, and fetal death.

The American Academy of Pediatrics recommends limiting exposure of children to pesticides and using safer alternatives:

The World Health Organization and the UN Environment Programme estimate that each year, 3 million workers in agriculture in the developing world experience severe poisoning from pesticides, about 18,000 of whom die. Owing to inadequate regulation and safety precautions, 99% of pesticide related deaths occur in developing countries that account for only 25% of pesticide usage. According to one study, as many as 25 million workers in developing countries may suffer mild pesticide poisoning yearly. There are several careers aside from agriculture that may also put individuals at risk of health effects from pesticide exposure including pet groomers, groundskeepers, and fumigators.

One study found pesticide self-poisoning the method of choice in one third of suicides worldwide, and recommended, among other things, more restrictions on the types of pesticides that are most harmful to humans.

A 2014 epidemiological review found associations between autism and exposure to certain pesticides, but noted that the available evidence was insufficient to conclude that the relationship was causal.

Environmental Effect

Pesticide use raises a number of environmental concerns. Over 98% of sprayed insecticides and 95% of herbicides reach a destination other than their target species, including non-target species, air, water and soil. Pesticide drift occurs when pesticides suspended in the air as particles are carried by wind to other areas, potentially contaminating them. Pesticides are one of the causes of water pollution, and some pesticides are persistent organic pollutants and contribute to soil contamination.

In addition, pesticide use reduces biodiversity, contributes to pollinator decline, destroys habitat (especially for birds), and threatens endangered species. Pests can develop a resistance to the pesticide (pesticide resistance), necessitating a new pesticide. Alternatively a greater dose of the pesticide can be used to counteract the resistance, although this will cause a worsening of the ambient pollution problem.

Since chlorinated hydrocarbon pesticides dissolve in fats and are not excreted, organisms tend to retain them almost indefinitely. Biological magnification is the process whereby these chlorinated hydrocarbons (pesticides) are more concentrated at each level of the food chain. Among marine animals, pesticide concentrations are higher in carnivorous fishes, and even more so in the fish-eating birds and mammals at the top of the ecological pyramid. Global distillation is the process whereby pesticides are transported from warmer to colder regions of the Earth, in particular the Poles and mountain tops. Pesticides that evaporate into the atmosphere at relatively high temperature can be carried considerable distances (thousands of kilometers) by the wind to an area of lower temperature, where they condense and are carried back to the ground in rain or snow.

In order to reduce negative impacts, it is desirable that pesticides be degradable or at least quickly deactivated in the environment. Such loss of activity or toxicity of pesticides is due to both innate chemical properties of the compounds and environmental processes or conditions. For example, the presence of halogens within a chemical structure often slows down degradation in an aerobic environment. Adsorption to soil may retard pesticide movement, but also may reduce bioavailability to microbial degraders.

Economics

Harm	Annual US cost
Public health	$1.1 billion
Pesticide resistance in pest	$1.5 billion
Crop losses caused by pesticides	$1.4 billion
Bird losses due to pesticides	$2.2 billion
Groundwater contamination	$2.0 billion
Other costs	$1.4 billion
Total costs	**$9.6 billion**

Human health and environmental cost from pesticides in the United States is estimated at $9.6 billion offset by about $40 billion in increased agricultural production:

Additional costs include the registration process and the cost of purchasing pesticides. The registration process can take several years to complete (there are 70 different types of field test) and can cost $50–70 million for a single pesticide. Annually the United States spends $10 billion on pesticides.

Alternatives

Alternatives to pesticides are available and include methods of cultivation, use of biological pest controls (such as pheromones and microbial pesticides), genetic engineering, and methods of interfering with insect breeding. Application of composted yard waste has also been used as a way of controlling pests. These methods are becoming increasingly popular and often are safer than traditional chemical pesticides. In addition, EPA is registering reduced-risk conventional pesticides in increasing numbers.

Cultivation practices include polyculture (growing multiple types of plants), crop rotation, planting crops in areas where the pests that damage them do not live, timing planting according to when pests will be least problematic, and use of trap crops that attract pests away from the real crop. In the U.S., farmers have had success controlling insects by spraying with hot water at a cost that is about the same as pesticide spraying.

Release of other organisms that fight the pest is another example of an alternative to pesticide use. These organisms can include natural predators or parasites of the pests. Biological pesticides based on entomopathogenic fungi, bacteria and viruses cause disease in the pest species can also be used.

Interfering with insects' reproduction can be accomplished by sterilizing males of the target species and releasing them, so that they mate with females but do not produce offspring. This technique was first used on the screwworm fly in 1958 and has since been used with the medfly, the tsetse fly, and the gypsy moth. However, this can be a costly, time consuming approach that only works on some types of insects.

Agroecology emphasize nutrient recycling, use of locally available and renewable resources, adaptation to local conditions, utilization of microenvironments, reliance on indigenous knowledge and yield maximization while maintaining soil productivity. Agroecology also emphasizes empowering people and local communities to contribute to development, and encouraging "multi-directional" communications rather than the conventional "top-down" method.

Push Pull Strategy

The term "push-pull" was established in 1987 as an approach for integrated pest management (IPM). This strategy uses a mixture of behavior-modifying stimuli to manipulate the distribution and abundance of insects. "Push" means the insects are repelled or deterred away from whatever resource that is being protected. "Pull" means that certain stimuli (semiochemical stimuli, pheromones, food additives, visual stimuli, genetically altered plants, etc.) are used to attract pests to trap crops where they will be killed. There are numerous different components involved in order to implement a Push-Pull Strategy in IPM.

Many case studies testing the effectiveness of the push-pull approach have been done across the world. The most successful push-pull strategy was developed in Africa for subsistence farming. Another successful case study was performed on the control of *Helicoverpa* in cotton crops in Australia. In Europe, the Middle East, and the United States, push-pull strategies were successfully used in the controlling of *Sitona lineatus* in bean fields.

Some advantages of using the push-pull method are less use of chemical or biological materials and better protection against insect habituation to this control method. Some disadvantages of the push-pull strategy is that if there is a lack of appropriate knowledge of behavioral and chemical ecology of the host-pest interactions then this method becomes unreliable. Furthermore, because the push-pull method is not a very popular method of IPM operational and registration costs are higher.

Effectiveness

Some evidence shows that alternatives to pesticides can be equally effective as the use of chemicals. For example, Sweden has halved its use of pesticides with hardly any reduction in crops. In Indonesia, farmers have reduced pesticide use on rice fields by 65% and experienced a 15% crop increase. A study of Maize fields in northern Florida found that the application of composted yard waste with high carbon to nitrogen ratio to agricultural fields was highly effective at reducing the population of plant-parasitic nematodes and increasing crop yield, with yield increases ranging from 10% to 212%; the observed effects were long-term, often not appearing until the third season of the study.

However, pesticide resistance is increasing. In the 1940s, U.S. farmers lost only 7% of their crops to pests. Since the 1980s, loss has increased to 13%, even though more pesticides are being used. Between 500 and 1,000 insect and weed species have developed pesticide resistance since 1945.

Types

Pesticides are often referred to according to the type of pest they control. Pesticides can also be considered as either biodegradable pesticides, which will be broken down by microbes and other living beings into harmless compounds, or persistent pesticides, which may take months or years before they are broken down: it was the persistence of DDT, for example, which led to its accumulation in the food chain and its killing of birds of prey at the top of the food chain. Another way to think about pesticides is to consider those that are chemical pesticides or are derived from a common source or production method.

Some examples of chemically-related pesticides are:

Organophosphate Pesticides

Organophosphates affect the nervous system by disrupting, acetylcholinesterase activity, the enzyme that regulates acetylcholine, a neurotransmitter. Most organophosphates are insecticides. They were developed during the early 19th century, but their effects on insects, which are similar to their effects on humans, were discovered in 1932. Some are very poisonous. However, they usually are not persistent in the environment.

Carbamate Pesticides

Carbamate pesticides affect the nervous system by disrupting an enzyme that regulates acetylcholine, a neurotransmitter. The enzyme effects are usually reversible. There are several subgroups within the carbamates.

Organochlorine Insecticides

They were commonly used in the past, but many have been removed from the market due to their health and environmental effects and their persistence (e.g., DDT, chlordane, and toxaphene).

Pyrethroid Pesticides

They were developed as a synthetic version of the naturally occurring pesticide pyrethrin, which is found in chrysanthemums. They have been modified to increase their stability in the environment. Some synthetic pyrethroids are toxic to the nervous system.

Sulfonylurea Herbicides

The following sulfonylureas have been commercialized for weed control: amidosulfuron, azimsulfuron, bensulfuron-methyl, chlorimuron-ethyl, ethoxysulfuron, flazasulfuron, flupyrsulfuron-methyl-sodium, halosulfuron-methyl, imazosulfuron, nicosulfuron, oxasulfuron, primisulfuron-methyl, pyrazosulfuron-ethyl, rimsulfuron, sulfometuron-methyl Sulfosulfuron, terbacil, bispyribac-sodium, cyclosulfamuron, and pyrithiobac-sodium. Nicosulfuron, triflusulfuron methyl, and chlorsulfuron are broad-spectrum herbicides that kill plants by inhibiting the enzyme acetolactate synthase. In the 1960s, more than 1 kg/ha (0.89 lb/acre) crop protection chemical was typically applied, while sulfonylureates allow as little as 1% as much material to achieve the same effect.

Biopesticides

Biopesticides are certain types of pesticides derived from such natural materials as animals, plants, bacteria, and certain minerals. For example, canola oil and baking soda have pesticidal applications and are considered biopesticides. Biopesticides fall into three major classes:

- Microbial pesticides which consist of bacteria, entomopathogenic fungi or viruses (and sometimes includes the metabolites that bacteria or fungi produce). Entomopathogenic nematodes are also often classed as microbial pesticides, even though they are multicellular.

- Biochemical pesticides or herbal pesticides are naturally occurring substances that control (or monitor in the case of pheromones) pests and microbial diseases.

- Plant-incorporated protectants (PIPs) have genetic material from other species incorporated into their genetic material (*i.e.* GM crops). Their use is controversial, especially in many European countries.

Classified by Type of Pest

Pesticides that are related to the type of pests are:

Type	Action
Algicides	Control algae in lakes, canals, swimming pools, water tanks, and other sites
Antifouling agents	Kill or repel organisms that attach to underwater surfaces, such as boat bottoms

Antimicrobials	Kill microorganisms (such as bacteria and viruses)
Attractants	Attract pests (for example, to lure an insect or rodent to a trap). (However, food is not considered a pesticide when used as an attractant.)
Biopesticides	Biopesticides are certain types of pesticides derived from such natural materials as animals, plants, bacteria, and certain minerals
Biocides	Kill microorganisms
Disinfectants and sanitizers	Kill or inactivate disease-producing microorganisms on inanimate objects
Fungicides	Kill fungi (including blights, mildews, molds, and rusts)
Fumigants	Produce gas or vapor intended to destroy pests in buildings or soil
Herbicides	Kill weeds and other plants that grow where they are not wanted
Insecticides	Kill insects and other arthropods
Miticides	Kill mites that feed on plants and animals
Microbial pesticides	Microorganisms that kill, inhibit, or out compete pests, including insects or other microorganisms
Molluscicides	Kill snails and slugs
Nematicides	Kill nematodes (microscopic, worm-like organisms that feed on plant roots)
Ovicides	Kill eggs of insects and mites
Pheromones	Biochemicals used to disrupt the mating behavior of insects
Repellents	Repel pests, including insects (such as mosquitoes) and birds
Rodenticides	Control mice and other rodents

Further Types of Pesticides

The term pesticide also include these substances:

Defoliants : Cause leaves or other foliage to drop from a plant, usually to facilitate harvest. Desiccants : Promote drying of living tissues, such as unwanted plant tops. Insect growth regulators : Disrupt the molting, maturity from pupal stage to adult, or other life processes of insects. Plant growth regulators : Substances (excluding fertilizers or other plant nutrients) that alter the expected growth, flowering, or reproduction rate of plants.

Regulation

International

In most countries, pesticides must be approved for sale and use by a government agency.

In Europe, recent EU legislation has been approved banning the use of highly toxic pesticides including those that are carcinogenic, mutagenic or toxic to reproduction, those that are endocrine-disrupting, and those that are persistent, bioaccumulative and toxic (PBT) or very persistent and very bioaccumulative (vPvB). Measures were approved to improve the general safety of pesticides across all EU member states.

Though pesticide regulations differ from country to country, pesticides, and products on which they were used are traded across international borders. To deal with inconsistencies in regulations among countries, delegates to a conference of the United Nations Food and Agriculture Organization adopted an International Code of Conduct on the Distribution and Use of Pesticides in 1985 to create voluntary standards of pesticide regulation for different countries. The Code was updated in 1998 and 2002. The FAO claims that the code has raised awareness about pesticide hazards and decreased the number of countries without restrictions on pesticide use.

Three other efforts to improve regulation of international pesticide trade are the United Nations London Guidelines for the Exchange of Information on Chemicals in International Trade and the United Nations Codex Alimentarius Commission. The former seeks to implement procedures for ensuring that prior informed consent exists between countries buying and selling pesticides, while the latter seeks to create uniform standards for maximum levels of pesticide residues among participating countries. Both initiatives operate on a voluntary basis.

Pesticides safety education and pesticide applicator regulation are designed to protect the public from pesticide misuse, but do not eliminate all misuse. Reducing the use of pesticides and choosing less toxic pesticides may reduce risks placed on society and the environment from pesticide use. Integrated pest management, the use of multiple approaches to control pests, is becoming widespread and has been used with success in countries such as Indonesia, China, Bangladesh, the U.S., Australia, and Mexico. IPM attempts to recognize the more widespread impacts of an action on an ecosystem, so that natural balances are not upset. New pesticides are being developed, including biological and botanical derivatives and alternatives that are thought to reduce health and environmental risks. In addition, applicators are being encouraged to consider alternative controls and adopt methods that reduce the use of chemical pesticides.

Pesticides can be created that are targeted to a specific pest's lifecycle, which can be environmentally more friendly. For example, potato cyst nematodes emerge from their protective cysts in response to a chemical excreted by potatoes; they feed on the potatoes and damage the crop. A similar chemical can be applied to fields early, before the potatoes are planted, causing the nematodes to emerge early and starve in the absence of potatoes.

United States

Preparation for an application of hazardous herbicide in USA.

In the United States, the Environmental Protection Agency (EPA) is responsible for regulating pesticides under the Federal Insecticide, Fungicide, and Rodenticide Act (FIFRA) and the Food Quality Protection Act (FQPA). Studies must be conducted to establish the conditions in which the material is safe to use and the effectiveness against the intended pest(s). The EPA regulates pesticides to ensure that these products do not pose adverse effects to humans or the environment. Pesticides produced before November 1984 continue to be reassessed in order to meet the current scientific and regulatory standards. All registered pesticides are reviewed every 15 years to ensure they meet the proper standards. During the registration process, a label is created. The label contains directions for proper use of the material in addition to safety restrictions. Based on acute toxicity, pesticides are assigned to a Toxicity Class.

Some pesticides are considered too hazardous for sale to the general public and are designated restricted use pesticides. Only certified applicators, who have passed an exam, may purchase or supervise the application of restricted use pesticides. Records of sales and use are required to be maintained and may be audited by government agencies charged with the enforcement of pesticide regulations. These records must be made available to employees and state or territorial environmental regulatory agencies.

The EPA regulates pesticides under two main acts, both of which amended by the Food Quality Protection Act of 1996. In addition to the EPA, the United States Department of Agriculture (USDA) and the United States Food and Drug Administration (FDA) set standards for the level of pesticide residue that is allowed on or in crops. The EPA looks at what the potential human health and environmental effects might be associated with the use of the pesticide.

In addition, the U.S. EPA uses the National Research Council's four-step process for human health risk assessment: (1) Hazard Identification, (2) Dose-Response Assessment, (3) Exposure Assessment, and (4) Risk Characterization.

Recently Kaua'i County (Hawai'i) passed Bill No. 2491 to add an article to Chapter 22 of the county's code relating to pesticides and GMOs. The bill strengthens protections of local communities in Kaua'i where many large pesticide companies test their products.

History

Since before 2000 BC, humans have utilized pesticides to protect their crops. The first known pesticide was elemental sulfur dusting used in ancient Sumer about 4,500 years ago in ancient Mesopotamia. The Rig Veda, which is about 4,000 years old, mentions the use of poisonous plants for pest control. By the 15th century, toxic chemicals such as arsenic, mercury, and lead were being applied to crops to kill pests. In the 17th century, nicotine sulfate was extracted from tobacco leaves for use as an insecticide. The 19th century saw the introduction of two more natural pesticides, pyrethrum, which is derived from chrysanthemums, and rotenone, which is derived from the roots of tropical vegetables. Until the 1950s, arsenic-based pesticides were dominant. Paul Müller discovered that DDT was a very effective insecticide. Organochlorines such as DDT were dominant, but they were replaced in the U.S. by organophosphates and carbamates by 1975. Since then, pyrethrin compounds have become the dominant insecticide. Herbicides became common in the 1960s, led by "triazine and other nitrogen-based compounds, carboxylic acids such as 2,4-dichlorophenoxyacetic acid, and glyphosate".

The first legislation providing federal authority for regulating pesticides was enacted in 1910; however, decades later during the 1940s manufacturers began to produce large amounts of synthetic pesticides and their use became widespread. Some sources consider the 1940s and 1950s to have been the start of the "pesticide era." Although the U.S. Environmental Protection Agency was established in 1970 and amendments to the pesticide law in 1972, pesticide use has increased 50-fold since 1950 and 2.3 million tonnes (2.5 million short tons) of industrial pesticides are now used each year. Seventy-five percent of all pesticides in the world are used in developed countries, but use in developing countries is increasing. A study of USA pesticide use trends through 1997 was published in 2003 by the National Science Foundation's Center for Integrated Pest Management.

In the 1960s, it was discovered that DDT was preventing many fish-eating birds from reproducing, which was a serious threat to biodiversity. Rachel Carson wrote the best-selling book *Silent Spring* about biological magnification. The agricultural use of DDT is now banned under the Stockholm Convention on Persistent Organic Pollutants, but it is still used in some developing nations to prevent malaria and other tropical diseases by spraying on interior walls to kill or repel mosquitoes.

Herbicide

Weeds controlled with herbicide

Herbicide(s), also commonly known as weedkillers, are chemical substances used to control unwanted plants. Selective herbicides control specific weed species, while leaving the desired crop relatively unharmed, while non-selective herbicides (sometimes called "total weedkillers" in commercial products) can be used to clear waste ground, industrial and construction sites, railways and railway embankments as they kill all plant material with which they come into contact. Apart from selective/non-selective, other important distinctions include *persistence* (also known as *residual action*: how long the product stays in place and remains active), *means of uptake* (whether it is absorbed by above-ground foliage only, through the roots, or by other means), and *mechanism of action* (how it works). Historically, products such as common salt and other metal salts were used as herbicides, however these have gradually fallen out of favor and in some countries a number of these are banned due to their persistence in soil, and toxicity and groundwater contamination concerns. Herbicides have also been used in warfare and conflict.

Modern herbicides are often synthetic mimics of natural plant hormones which interfere with growth of the target plants. The term organic herbicide has come to mean herbicides intended for organic farming; these are often less efficient and more costly than synthetic herbicides and are based on natural materials. Some plants also produce their own natural herbicides, such as the genus *Juglans* (walnuts), or the tree of heaven; such action of natural herbicides, and other related chemical interactions, is called allelopathy. Due to herbicide resistance - a major concern in agriculture - a number of products also combine herbicides with different means of action.

In the US in 2007, about 83% of all herbicide usage, determined by weight applied, was in agriculture. In 2007, world pesticide expenditures totaled about $39.4 billion; herbicides were about 40% of those sales and constituted the biggest portion, followed by insecticides, fungicides, and other types. Smaller quantities are used in forestry, pasture systems, and management of areas set aside as wildlife habitat.

History

Prior to the widespread use of chemical herbicides, cultural controls, such as altering soil pH, salinity, or fertility levels, were used to control weeds. Mechanical control (including tillage) was also (and still is) used to control weeds.

First Herbicides

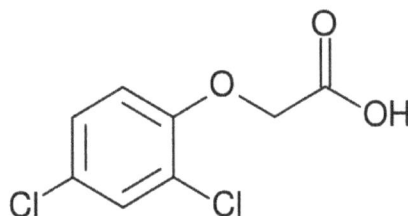

2,4-D, the first chemical herbicide, was discovered during the Second World War.

Although research into chemical herbicides began in the early 20th century, the first major breakthrough was the result of research conducted in both the UK and the US during the Second World War into the potential use of agents as biological weapons. The first modern herbicide, 2,4-D, was first discovered and synthesized by W. G. Templeman at Imperial Chemical Industries. In 1940, he showed that "Growth substances applied appropriately would kill certain broad-leaved weeds in cereals without harming the crops." By 1941, his team succeeded in synthesizing the chemical. In the same year, Pokorny in the US achieved this as well.

Independently, a team under Juda Hirsch Quastel, working at the Rothamsted Experimental Station made the same discovery. Quastel was tasked by the Agricultural Research Council (ARC) to discover methods for improving crop yield. By analyzing soil as a dynamic system, rather than an inert substance, he was able to apply techniques such as perfusion. Quastel was able to quantify the influence of various plant hormones, inhibitors and other chemicals on the activity of microorganisms in the soil and assess their direct impact on plant growth. While the full work of the unit remained secret, certain discoveries were developed for commercial use after the war, including the 2,4-D compound.

When it was commercially released in 1946, it triggered a worldwide revolution in agricultural output and became the first successful selective herbicide. It allowed for greatly enhanced weed control in wheat, maize (corn), rice, and similar cereal grass crops, because it kills dicots (broad-leaf plants), but not most monocots (grasses). The low cost of 2,4-D has led to continued usage today, and it remains one of the most commonly used herbicides in the world. Like other acid herbicides, current formulations use either an amine salt (often trimethylamine) or one of many esters of the parent compound. These are easier to handle than the acid.

Further Discoveries

The triazine family of herbicides, which includes atrazine, were introduced in the 1950s; they have the current distinction of being the herbicide family of greatest concern regarding ground-water contamination. Atrazine does not break down readily (within a few weeks) after being applied to soils of above neutral pH. Under alkaline soil conditions, atrazine may be carried into the soil profile as far as the water table by soil water following rainfall causing the aforementioned contamination. Atrazine is thus said to have "carryover", a generally undesirable property for herbicides.

Glyphosate (Roundup) was introduced in 1974 for nonselective weed control. Following the development of glyphosate-resistant crop plants, it is now used very extensively for selective weed control in growing crops. The pairing of the herbicide with the resistant seed contributed to the consolidation of the seed and chemistry industry in the late 1990s.

Many modern chemical herbicides used in agriculture and gardening are specifically formulated to decompose within a short period after application. This is desirable, as it allows crops and plants to be planted afterwards, which could otherwise be affected by the herbicide. However, herbicides with low residual activity (i.e., that decompose quickly) often do not provide season-long weed control and do not ensure that weed roots are killed beneath construction and paving (and cannot emerge destructively in years to come), therefore there remains a role for weedkiller with high levels of persistence in the soil.

Terminology

Herbicides are classified/grouped in various ways e.g. according to the activity, timing of application, method of application, mechanism of action, chemical family. This gives rise to a considerable level of terminology related to herbicides and their use.

Intended Outcome

- Control is the destruction of unwanted weeds, or the damage of them to the point where they are no longer competitive with the crop.

- Suppression is incomplete control still providing some economic benefit, such as reduced competition with the crop.

- Crop safety, for selective herbicides, is the relative absence of damage or stress to the crop. Most selective herbicides cause some visible stress to crop plants.

- Defoliant, similar to herbicides, but designed to remove foliage (leaves) rather than kill the plant.

Selectivity (All Plants or Specific Plants)

- Selective herbicides: They control or suppress certain plants without affecting the growth of other plants species. Selectivity may be due to translocation, differential absorption, physical (morphological) or physiological differences between plant species. 2,4-D, mecoprop, dicamba control many broadleaf weeds but remain ineffective against turfgrasses.

- Non-selective herbicides: These herbicides are not specific in acting against certain plant species and control all plant material with which they come into contact. They are used to clear industrial sites, waste ground, railways and railway embankments. Paraquat, glufosinate, glyphosate are non-selective herbicides.

Timing of Application

- Preplant: Preplant herbicides are nonselective herbicides applied to soil before planting. Some preplant herbicides may be mechanically incorporated into the soil. The objective for incorporation is to prevent dissipation through photodecomposition and/or volatility. The herbicides kill weeds as they grow through the herbicide treated zone. Volatile herbicides have to be incorporated into the soil before planting the pasture. Agricultural crops grown in soil treated with a preplant herbicide include tomatoes, corn, soybeans and strawberries. Soil fumigants like metam-sodium and dazomet are in use as preplant herbicides.

- Preemergence: Preemergence herbicides are applied before the weed seedlings emerge through the soil surface. Herbicides do not prevent weeds from germinating but they kill weeds as they grow through the herbicide treated zone by affecting the cell division in the emerging seedling. Dithopyr and pendimethalin are preemergence herbicides. Weeds that have already emerged before application or activation are not affected by pre-herbicides as their primary growing point escapes the treatment.

- Postemergence: These herbicides are applied after weed seedlings have emerged through the soil surface. They can be foliar or root absorbed, selective or nonselective, contact or systemic. Application of these herbicides is avoided during rain because the problem of being washed off to the soil makes it ineffective. 2,4-D is a selective, systemic, foliar absorbed postemergence herbicide.

Method of Application

- Soil applied: Herbicides applied to the soil are usually taken up by the root or shoot of the emerging seedlings and are used as preplant or preemergence treatment. Several factors influence the effectiveness of soil-applied herbicides. Weeds absorb herbicides by both passive and active mechanism. Herbicide adsorption to soil colloids or organic matter often reduces its amount available for weed absorption. Positioning of herbicide in correct layer of soil is very important, which can be achieved mechanically and by rainfall. Herbicides on the soil surface are subjected to several processes that reduce their availability. Volatility and photolysis are two common processes that reduce the availability of herbicides. Many soil applied herbicides are absorbed through plant shoots while they are still underground leading to their death or injury. EPTC and trifluralin are soil applied herbicides.

- Foliar applied: These are applied to portion of the plant above the ground and are absorbed by exposed tissues. These are generally postemergence herbicides and can either be translocated (systemic) throughout the plant or remain at specific site (contact). External barriers of plants like cuticle, waxes, cell wall etc. affect herbicide absorption and action. Glyphosate, 2,4-D and dicamba are foliar applied herbicide.

Persistence

- Residual activity: A herbicide is described as having low residual activity if it is neutralized within a short time of application (within a few weeks or months) - typically this is due to rainfall, or by reactions in the soil. A herbicide described as having high residual activity will remains potent for a long term in the soil. For some compounds, the residual activity can leave the ground almost permanently barren.

Mechanism of Action

Herbicides are often classified according to their site of action, because as a general rule, herbicides within the same site of action class will produce similar symptoms on susceptible plants. Classification based on site of action of herbicide is comparatively better as herbicide resistance management can be handled more properly and effectively. Classification by mechanism of action (MOA) indicates the first enzyme, protein, or biochemical step affected in the plant following application.

List of Mechanisms Found in Modern Herbicides

- ACCase inhibitors compounds kill grasses. Acetyl coenzyme A carboxylase (ACCase) is part of the first step of lipid synthesis. Thus, ACCase inhibitors affect cell membrane production in the meristems of the grass plant. The ACCases of grasses are sensitive to these herbicides, whereas the ACCases of dicot plants are not.

- ALS inhibitors: the acetolactate synthase (ALS) enzyme (also known as acetohydroxyacid synthase, or AHAS) is the first step in the synthesis of the branched-chain amino acids (valine, leucine, and isoleucine). These herbicides slowly starve affected plants of these amino acids, which eventually leads to inhibition of DNA synthesis. They affect grasses and dicots alike. The ALS inhibitor family includes various sulfonylureas (such as Flazasulfuron and Metsulfuron-methyl), imidazolinones, triazolopyrimidines, pyrimidinyl oxybenzoates, and sulfonylamino carbonyl triazolinones. The ALS biological pathway exists only in plants and not animals, thus making the ALS-inhibitors among the safest herbicides.

- EPSPS inhibitors: The enolpyruvylshikimate 3-phosphate synthase enzyme EPSPS is used in the synthesis of the amino acids tryptophan, phenylalanine and tyrosine. They affect grasses and dicots alike. Glyphosate (Roundup) is a systemic EPSPS inhibitor inactivated by soil contact.

- Synthetic auxins inaugurated the era of organic herbicides. They were discovered in the 1940s after a long study of the plant growth regulator auxin. Synthetic auxins mimic this plant hormone. They have several points of action on the cell membrane, and are effective in the control of dicot plants. 2,4-D is a synthetic auxin herbicide.

- Photosystem II inhibitors reduce electron flow from water to NADPH2+ at the photochemical step in photosynthesis. They bind to the Qb site on the D1 protein, and prevent quinone from binding to this site. Therefore, this group of compounds causes electrons to accumulate on chlorophyll molecules. As a consequence, oxidation reactions in excess of those normally tolerated by the cell occur, and the plant dies. The triazine herbicides (including atrazine) and urea derivatives (diuron) are photosystem II inhibitors.

- Photosystem I inhibitors steal electrons from the normal pathway through FeS to Fdx to NADP leading to direct discharge of electrons on oxygen. As a result, reactive oxygen species are produced and oxidation reactions in excess of those normally tolerated by the cell occur, leading to plant death. Bipyridinium herbicides (such as diquat and paraquat) inhibit the Fe-S – Fdx step of that chain, while diphenyl ether herbicides (such as nitrofen, nitrofluorfen, and acifluorfen) inhibit the Fdx – NADP step.

- HPPD inhibitors inhibit 4-Hydroxyphenylpyruvate dioxygenase, which are involved in tyrosine breakdown. Tyrosine breakdown products are used by plants to make carotenoids, which protect chlorophyll in plants from being destroyed by sunlight. If this happens, the plants turn white due to complete loss of chlorophyll, and the plants die. Mesotrione and sulcotrione are herbicides in this class; a drug, nitisinone, was discovered in the course of developing this class of herbicides.

Herbicide Group (Labeling)

One of the most important methods for preventing, delaying, or managing resistance is to reduce the reliance on a single herbicide mode of action. To do this, farmers must know the mode of action for the herbicides they intend to use, but the relatively complex nature of plant biochemistry makes this difficult to determine. Attempts were made to simplify the understanding of herbicide mode of action by developing a classification system that grouped herbicides by mode of action. Eventually the Herbicide Resistance Action Committee (HRAC) and the Weed Science Society of America (WSSA) developed a classification system. The WSSA and HRAC systems differ in the group designation. Groups in the WSSA and the HRAC systems are designated by numbers and letters, respectively. The goal for adding the "Group" classification and mode of action to the herbicide product label is to provide a simple and practical approach to deliver the information to users. This information will make it easier to develop educational material that is consistent and effective. It should increase user's awareness of herbicide mode of action and provide more accurate recommendations for resistance management. Another goal is to make it easier for users to keep records on which herbicide mode of actions are being used on a particular field from year to year.

Chemical Family

Detailed investigations on chemical structure of the active ingredients of the registered herbicides showed that some moieties (moiety is a part of a molecule that may include either whole functional groups or parts of functional groups as substructures; a functional group has similar chemical properties whenever it occurs in different compounds) have the same mechanisms of action. According to Forouzesh *et al.* 2015, these moieties have been assigned to the names of chemical families and active ingredients are then classified within the chemical families accordingly. Knowing about herbicide chemical family grouping could serve as a short-term strategy for managing resistance to site of action.

Use and Application

Herbicides being sprayed from the spray arms of a tractor in North Dakota.

Most herbicides are applied as water-based sprays using ground equipment. Ground equipment varies in design, but large areas can be sprayed using self-propelled sprayers equipped with long booms, of 60 to 120 feet (18 to 37 m) with spray nozzles spaced every 20–30 inches (510–760 mm) apart. Towed, handheld, and even horse-drawn sprayers are also used. On large areas, herbicides may also at times be applied aerially using helicopters or airplanes, or through irrigation systems (known as chemigation).

A further method of herbicide application developed around 2010, involves ridding the soil of its active weed seed bank rather than just killing the weed. This can successfully treat annual plants but not perennials. Researchers at the Agricultural Research Service found that the application of herbicides to fields late in the weeds' growing season greatly reduces their seed production, and therefore fewer weeds will return the following season. Because most weeds are annuals, their seeds will only survive in soil for a year or two, so this method will be able to destroy such weeds after a few years of herbicide application.

Weed-wiping may also be used, where a wick wetted with herbicide is suspended from a boom and dragged or rolled across the tops of the taller weed plants. This allows treatment of taller grassland weeds by direct contact without affecting related but desirable shorter plants in the grassland sward beneath. The method has the benefit of avoiding spray drift. In Wales, a scheme offering free weed-wiper hire was launched in 2015 in an effort to reduce the levels of MCPA in water courses.

Misuse and Misapplication

Herbicide volatilisation or spray drift may result in herbicide affecting neighboring fields or plants, particularly in windy conditions. Sometimes, the wrong field or plants may be sprayed due to error.

Use politically, Militarily, and in Conflict

Health and Environmental Effects

Herbicides have widely variable toxicity in addition to acute toxicity from occupational exposure levels.

Some herbicides cause a range of health effects ranging from skin rashes to death. The pathway of attack can arise from intentional or unintentional direct consumption, improper application resulting in the herbicide coming into direct contact with people or wildlife, inhalation of aerial sprays, or food consumption prior to the labeled preharvest interval. Under some conditions, certain herbicides can be transported via leaching or surface runoff to contaminate groundwater or distant surface water sources. Generally, the conditions that promote herbicide transport include intense storm events (particularly shortly after application) and soils with limited capacity to adsorb or retain the herbicides. Herbicide properties that increase likelihood of transport include persistence (resistance to degradation) and high water solubility.

Phenoxy herbicides are often contaminated with dioxins such as TCDD; research has suggested such contamination results in a small rise in cancer risk after occupational exposure to these herbicides. Triazine exposure has been implicated in a likely relationship to increased risk of breast cancer, although a causal relationship remains unclear.

Herbicide manufacturers have at times made false or misleading claims about the safety of their products. Chemical manufacturer Monsanto Company agreed to change its advertising after pressure from New York attorney general Dennis Vacco; Vacco complained about misleading claims that its spray-on glyphosate-based herbicides, including Roundup, were safer than table salt and "practically non-toxic" to mammals, birds, and fish (though proof that this was ever said is hard to find). Roundup is toxic and has resulted in death after being ingested in quantities ranging from 85 to 200 ml, although it has also been ingested in quantities as large as 500 ml with only mild or moderate symptoms. The manufacturer of Tordon 101 (Dow AgroSciences, owned by the Dow Chemical Company) has claimed Tordon 101 has no effects on animals and insects, in spite of evidence of strong carcinogenic activity of the active ingredient Picloram in studies on rats.

The risk of Parkinson's disease has been shown to increase with occupational exposure to herbicides and pesticides. The herbicide paraquat is suspected to be one such factor.

All commercially sold, organic and nonorganic herbicides must be extensively tested prior to approval for sale and labeling by the Environmental Protection Agency. However, because of the large number of herbicides in use, concern regarding health effects is significant. In addition to health effects caused by herbicides themselves, commercial herbicide mixtures often contain other chemicals, including inactive ingredients, which have negative impacts on human health.

Ecological Effects

Commercial herbicide use generally has negative impacts on bird populations, although the impacts are highly variable and often require field studies to predict accurately. Laboratory studies have at times overestimated negative impacts on birds due to toxicity, predicting serious problems that were not observed in the field. Most observed effects are due not to toxicity, but to habitat changes and the decreases in abundance of species on which birds rely for food or shelter. Herbicide use in silviculture, used to favor certain types of growth following clearcutting, can cause significant drops in bird populations. Even when herbicides which have low toxicity to birds are used, they decrease the abundance of many types of vegetation on which the birds rely. Herbicide use in agriculture in Britain has been linked to a decline in seed-eating bird species which rely on the weeds killed by the herbicides. Heavy use of herbicides in neotropical agricultural areas has been one of many factors implicated in limiting the usefulness of such agricultural land for wintering migratory birds.

Frog populations may be affected negatively by the use of herbicides as well. While some studies have shown that atrazine may be a teratogen, causing demasculinization in male frogs, the U.S. Environmental Protection Agency (EPA) and its independent Scientific Advisory Panel (SAP) examined all available studies on this topic and concluded that "atrazine does not adversely affect amphibian gonadal development based on a review of laboratory and field studies."

Scientific Uncertainty of Full Extent of Herbicide Effects

The health and environmental effects of many herbicides is unknown, and even the scientific community often disagrees on the risk. For example, a 1995 panel of 13 scientists reviewing studies on the carcinogenicity of 2,4-D had divided opinions on the likelihood 2,4-D causes cancer in humans. As of 1992, studies on phenoxy herbicides were too few to accurately assess the risk of many types of cancer from these herbicides, even though evidence was stronger that exposure to these herbicides is associated with increased risk of soft tissue sarcoma and non-Hodgkin lymphoma. Furthermore, there is some suggestion that herbicides can play a role in sex reversal of certain organisms that experience temperature-dependent sex determination, which could theoretically alter sex ratios.

Resistance

Weed resistance to herbicides has become a major concern in crop production worldwide. Resistance to herbicides is often attributed to lack of rotational programmes of herbicides and to continuous applications of herbicides with the same sites of action. Thus, a true understanding of the sites of action of herbicides is essential for strategic planning of herbicide-based weed control.

Plants have developed resistance to atrazine and to ALS-inhibitors, and more recently, to glyphosate herbicides. Marestail is one weed that has developed glyphosate resistance. Glyphosate-resistant weeds are present in the vast majority of soybean, cotton and corn farms in some U.S. states. Weeds that can resist multiple other herbicides are spreading. Few new herbicides are near commercialization, and none with a molecular mode of action for which there is no resistance. Because most herbicides could not kill all weeds, farmers rotated crops and herbicides to stop resistant weeds. During its initial years, glyphosate was not subject to resistance and allowed farmers to reduce the use of rotation.

A family of weeds that includes waterhemp (Amaranthus rudis) is the largest concern. A 2008-9 survey of 144 populations of waterhemp in 41 Missouri counties revealed glyphosate resistance in 69%. Weeds from some 500 sites throughout Iowa in 2011 and 2012 revealed glyphosate resistance in approximately 64% of waterhemp samples. The use of other killers to target "residual" weeds has become common, and may be sufficient to have stopped the spread of resistance From 2005 through 2010 researchers discovered 13 different weed species that had developed resistance to glyphosate. But since then only two more have been discovered. Weeds resistant to multiple herbicides with completely different biological action modes are on the rise. In Missouri, 43% of samples were resistant to two different herbicides; 6% resisted three; and 0.5% resisted four. In Iowa 89% of waterhemp samples resist two or more herbicides, 25% resist three, and 10% resist five.

For southern cotton, herbicide costs has climbed from between $50 and $75 per hectare a few years ago to about $370 per hectare in 2013. Resistance is contributing to a massive shift away from growing cotton; over the past few years, the area planted with cotton has declined by 70% in Arkansas and by 60% in Tennessee. For soybeans in Illinois, costs have risen from about $25 to $160 per hectare.

Dow, Bayer CropScience, Syngenta and Monsanto are all developing seed varieties resistant to herbicides other than glyphosate, which will make it easier for farmers to use alternative weed killers. Even though weeds have already evolved some resistance to those herbicides, Powles says the new seed-and-herbicide combos should work well if used with proper rotation.

Biochemistry of Resistance

Resistance to herbicides can be based on one of the following biochemical mechanisms:

- Target-site resistance: This is due to a reduced (or even lost) ability of the herbicide to bind to its target protein. The effect usually relates to an enzyme with a crucial function in a metabolic pathway, or to a component of an electron-transport system. Target-site resistance may also be caused by an overexpression of the target enzyme (via gene amplification or changes in a gene promoter).

- Non-target-site resistance: This is caused by mechanisms that reduce the amount of herbicidal active compound reaching the target site. One important mechanism is an enhanced metabolic detoxification of the herbicide in the weed, which leads to insufficient amounts of the active substance reaching the target site. A reduced uptake and translocation, or sequestration of the herbicide, may also result in an insufficient herbicide transport to the target site.

- Cross-resistance: In this case, a single resistance mechanism causes resistance to several herbicides. The term target-site cross-resistance is used when the herbicides bind to the same target site, whereas non-target-site cross-resistance is due to a single non-target-site mechanism (e.g., enhanced metabolic detoxification) that entails resistance across herbicides with different sites of action.

- Multiple resistance: In this situation, two or more resistance mechanisms are present within individual plants, or within a plant population.

Resistance Management

Worldwide experience has been that farmers tend to do little to prevent herbicide resistance developing, and only take action when it is a problem on their own farm or neighbor's. Careful observation is important so that any reduction in herbicide efficacy can be detected. This may indicate evolving resistance. It is vital that resistance is detected at an early stage as if it becomes an acute, whole-farm problem, options are more limited and greater expense is almost inevitable. Table 1 lists factors which enable the risk of resistance to be assessed. An essential pre-requisite for confirmation of resistance is a good diagnostic test. Ideally this should be rapid, accurate, cheap and accessible. Many diagnostic tests have been developed, including glasshouse pot assays, petri dish assays and chlorophyll fluorescence. A key component of such tests is that the response of the suspect population to a herbicide can be compared with that of known susceptible and resistant standards under controlled conditions. Most cases of herbicide resistance are a consequence of the repeated use of herbicides, often in association with crop monoculture and reduced cultivation practices. It is necessary, therefore, to modify these practices in order to prevent or delay the onset of resistance or to control existing resistant populations. A key objective should be the reduction in selection pressure. An integrated weed management (IWM) approach is required, in which as many tactics as possible are used to combat weeds. In this way, less reliance is placed on herbicides and so selection pressure should be reduced.

Optimising herbicide input to the economic threshold level should avoid the unnecessary use of herbicides and reduce selection pressure. Herbicides should be used to their greatest potential by ensuring that the timing, dose, application method, soil and climatic conditions are optimal for good activity. In the UK, partially resistant grass weeds such as *Alopecurus myosuroides* (black-grass) and *Avena* spp. (wild oat) can often be controlled adequately when herbicides are applied at the 2-3 leaf stage, whereas later applications at the 2-3 tiller stage can fail badly. Patch spraying, or applying herbicide to only the badly infested areas of fields, is another means of reducing total herbicide use.

Table 1. Agronomic factors influencing the risk of herbicide resistance development

Factor	Low risk	High risk
Cropping system	Good rotation	Crop monoculture
Cultivation system	Annual ploughing	Continuous minimum tillage
Weed control	Cultural only	Herbicide only
Herbicide use	Many modes of action	Single modes of action
Control in previous years	Excellent	Poor
Weed infestation	Low	High
Resistance in vicinity	Unknown	Common

Approaches to Treating Resistant Weeds

Alternative Herbicides

When resistance is first suspected or confirmed, the efficacy of alternatives is likely to be the first consideration. The use of alternative herbicides which remain effective on resistant populations can be a successful strategy, at least in the short term. The effectiveness of alternative herbicides will be highly dependent on the extent of cross-resistance. If there is resistance to a single group of herbicides, then the use of herbicides from other groups may provide a simple and effective solution, at least in the short term. For example, many triazine-resistant weeds have been readily controlled by the use of alternative herbicides such as dicamba or glyphosate. If resistance extends to more than one herbicide group, then choices are more limited. It should not be assumed that resistance will automatically extend to all herbicides with the same mode of action, although it is wise to assume this until proved otherwise. In many weeds the degree of cross-resistance between the five groups of ALS inhibitors varies considerably. Much will depend on the resistance mechanisms present, and it should not be assumed that these will necessarily be the same in different populations of the same species. These differences are due, at least in part, to the existence of different mutations conferring target site resistance. Consequently, selection for different mutations may result in different patterns of cross-resistance. Enhanced metabolism can affect even closely related herbicides to differing degrees. For example, populations of *Alopecurus myosuroides* (blackgrass) with an enhanced metabolism mechanism show resistance to pendimethalin but not to trifluralin, despite both being dinitroanilines. This is due to differences in the vulnerability of these two herbicides to oxidative metabolism. Consequently, care is needed when trying to predict the efficacy of alternative herbicides.

Mixtures and Sequences

The use of two or more herbicides which have differing modes of action can reduce the selection for resistant genotypes. Ideally, each component in a mixture should:

- Be active at different target sites
- Have a high level of efficacy
- Be detoxified by different biochemical pathways
- Have similar persistence in the soil (if it is a residual herbicide)
- Exert negative cross-resistance
- Synergise the activity of the other component

No mixture is likely to have all these attributes, but the first two listed are the most important. There is a risk that mixtures will select for resistance to both components in the longer term. One practical advantage of sequences of two herbicides compared with mixtures is that a better appraisal of the efficacy of each herbicide component is possible, provided that sufficient time elapses between each application. A disadvantage with sequences is that two separate applications have to be made and it is possible that the later application will be less effective on weeds surviving the first application. If these are resistant, then the second herbicide in the sequence may increase selection for resistant individuals by killing the susceptible plants which were damaged but not killed by the first application, but allowing the larger, less affected, resistant plants to survive. This has been cited as one reason why ALS-resistant *Stellaria media* has evolved in Scotland recently (2000), despite the regular use of a sequence incorporating mecoprop, a herbicide with a different mode of action.

Herbicide Rotations

Rotation of herbicides from different chemical groups in successive years should reduce selection for resistance. This is a key element in most resistance prevention programmes. The value of this approach depends on the extent of cross-resistance, and whether multiple resistance occurs owing to the presence of several different resistance mechanisms. A practical problem can be the lack of awareness by farmers of the different groups of herbicides that exist. In Australia a scheme has been introduced in which identifying letters are included on the product label as a means of enabling farmers to distinguish products with different modes of action.

Farming Practices and Resistance: A Case Study

Herbicide resistance became a critical problem in Australian agriculture, after many Australian sheep farmers began to exclusively grow wheat in their pastures in the 1970s. Introduced varieties of ryegrass, while good for grazing sheep, compete intensely with wheat. Ryegrasses produce so many seeds that, if left unchecked, they can completely choke a field. Herbicides provided excellent control, while reducing soil disrupting because of less need to plough. Within little more than a decade, ryegrass and other weeds began to develop resistance. In response Australian farmers changed methods. By 1983, patches of ryegrass had become immune to Hoegrass, a family of herbicides that inhibit an enzyme called acetyl coenzyme A carboxylase.

Ryegrass populations were large, and had substantial genetic diversity, because farmers had planted many varieties. Ryegrass is cross-pollinated by wind, so genes shuffle frequently. To control its distribution farmers sprayed inexpensive Hoegrass, creating selection pressure. In addition, farmers sometimes diluted the herbicide in order to save money, which allowed some plants to survive application. When resistance appeared farmers turned to a group of herbicides that block acetolactate synthase. Once again, ryegrass in Australia evolved a kind of "cross-resistance" that allowed it to rapidly break down a variety of herbicides. Four classes of herbicides become ineffective within a few years. In 2013 only two herbicide classes, called Photosystem II and long-chain fatty acid inhibitors, were effective against ryegrass.

List of Common Herbicides

Synthetic Herbicides

- 2,4-D is a broadleaf herbicide in the phenoxy group used in turf and no-till field crop production. Now, it is mainly used in a blend with other herbicides to allow lower rates of herbicides to be used; it is the most widely used herbicide in the world, and third most commonly used in the United States. It is an example of synthetic auxin (plant hormone).

- Aminopyralid is a broadleaf herbicide in the pyridine group, used to control weeds on grassland, such as docks, thistles and nettles. It is notorious for its ability to persist in compost.

- Atrazine, a triazine herbicide, is used in corn and sorghum for control of broadleaf weeds and grasses. Still used because of its low cost and because it works well on a broad spectrum of weeds common in the US corn belt, atrazine is commonly used with other herbicides to reduce the overall rate of atrazine and to lower the potential for groundwater contamination; it is a photosystem II inhibitor.

- Clopyralid is a broadleaf herbicide in the pyridine group, used mainly in turf, rangeland, and for control of noxious thistles. Notorious for its ability to persist in compost, it is another example of synthetic auxin.

- Dicamba, a postemergent broadleaf herbicide with some soil activity, is used on turf and field corn. It is another example of a synthetic auxin.

- Glufosinate ammonium, a broad-spectrum contact herbicide, is used to control weeds after the crop emerges or for total vegetation control on land not used for cultivation.

- Fluazifop (Fuselade Forte), a post emergence, foliar absorbed, translocated grass-selective herbicide with little residual action. It is used on a very wide range of broad leaved crops for control of annual and perennial grasses.

- Fluroxypyr, a systemic, selective herbicide, is used for the control of broad-leaved weeds in small grain cereals, maize, pastures, rangeland and turf. It is a synthetic auxin. In cereal growing, fluroxypyr's key importance is control of cleavers, *Galium aparine*. Other key broadleaf weeds are also controlled.

- Glyphosate, a systemic nonselective herbicide, is used in no-till burndown and for weed control in crops genetically modified to resist its effects. It is an example of an EPSPs inhibitor.

- Imazapyr a nonselective herbicide, is used for the control of a broad range of weeds, including terrestrial annual and perennial grasses and broadleaf herbs, woody species, and riparian and emergent aquatic species.

- Imazapic, a selective herbicide for both the pre- and postemergent control of some annual and perennial grasses and some broadleaf weeds, kills plants by inhibiting the production of branched chain amino acids (valine, leucine, and isoleucine), which are necessary for protein synthesis and cell growth.

- Imazamox, an imidazolinone manufactured by BASF for postemergence application that is an acetolactate synthase (ALS) inhibitor. Sold under trade names Raptor, Beyond, and Clearcast.

- Linuron is a nonselective herbicide used in the control of grasses and broadleaf weeds. It works by inhibiting photosynthesis.

- MCPA (2-methyl-4-chlorophenoxyacetic acid) is a phenoxy herbicide selective for broadleaf plants and widely used in cereals and pasture.

- Metolachlor is a pre-emergent herbicide widely used for control of annual grasses in corn and sorghum; it has displaced some of the atrazine in these uses.

- Paraquat is a nonselective contact herbicide used for no-till burndown and in aerial destruction of marijuana and coca plantings. It is more acutely toxic to people than any other herbicide in widespread commercial use.

- Pendimethalin, a pre-emergent herbicide, is widely used to control annual grasses and some broad-leaf weeds in a wide range of crops, including corn, soybeans, wheat, cotton, many tree and vine crops, and many turfgrass species.

- Picloram, a pyridine herbicide, mainly is used to control unwanted trees in pastures and edges of fields. It is another synthetic auxin.

- Sodium chlorate *(disused/banned in some countries)*, a nonselective herbicide, is considered phytotoxic to all green plant parts. It can also kill through root absorption.

- Triclopyr, a systemic, foliar herbicide in the pyridine group, is used to control broadleaf weeds while leaving grasses and conifers unaffected.

- Several sulfonylureas, including Flazasulfuron and Metsulfuron-methyl, which act as ALS inhibitors and in some cases are taken up from the soil via the roots.

Organic Herbicides

Recently, the term "organic" has come to imply products used in organic farming. Under this definition, an organic herbicide is one that can be used in a farming enterprise that has been classified as organic. Commercially sold organic herbicides are expensive and may not be affordable for commercial farming. Depending on the application, they may be less effective than synthetic herbicides and are generally used along with cultural and mechanical weed control practices.

Homemade organic herbicides include:

- Corn gluten meal (CGM) is a natural pre-emergence weed control used in turfgrass, which reduces germination of many broadleaf and grass weeds.

- Vinegar is effective for 5–20% solutions of acetic acid, with higher concentrations most effective, but it mainly destroys surface growth, so respraying to treat regrowth is needed. Resistant plants generally succumb when weakened by respraying.

- Steam has been applied commercially, but is now considered uneconomical and inadequate. It controls surface growth but not underground growth and so respraying to treat regrowth of perennials is needed.

- Flame is considered more effective than steam, but suffers from the same difficulties.

- D-limonene (citrus oil) is a natural degreasing agent that strips the waxy skin or cuticle from weeds, causing dehydration and ultimately death.

- Saltwater or salt applied in appropriate strengths to the rootzone will kill most plants.

- Monocerin produced by certain fungi will kill certain weeds such as Johnson grass.

Of Historical Interest and Other

- 2,4,5-Trichlorophenoxyacetic acid (2,4,5-T) was a widely used broadleaf herbicide until being phased out starting in the late 1970s. While 2,4,5-T itself is of only moderate toxicity, the manufacturing process for 2,4,5-T contaminates this chemical with trace amounts of 2,3,7,8-tetrachlorodibenzo-p-dioxin (TCDD). TCDD is extremely toxic to humans. With proper temperature control during production of 2,4,5-T, TCDD levels can be held to about .005 ppm. Before the TCDD risk was well understood, early production facilities lacked proper temperature controls. Individual batches tested later were found to have as much as 60 ppm of TCDD. 2,4,5-T was withdrawn from use in the USA in 1983, at a time of heightened public sensitivity about chemical hazards in the environment. Public concern about dioxins was high, and production and use of other (non-herbicide) chemicals potentially containing TCDD contamination was also withdrawn. These included penta-chlorophenol (a wood preservative) and PCBs (mainly used as stabilizing agents in transformer oil). Some feel that the 2,4,5-T withdrawal was not based on sound science. 2,4,5-T has since largely been replaced by dicamba and triclopyr.

- Agent Orange was a herbicide blend used by the British military during the Malayan Emergency and the U.S. military during the Vietnam War between January 1965 and April 1970 as a defoliant. It was a 50/50 mixture of the *n*-butyl esters of 2,4,5-T and 2,4-D. Because of TCDD contamination in the 2,4,5-T component, it has been blamed for serious illnesses in many people who were exposed to it. However, research on populations exposed to its dioxin contaminant have been inconsistent and inconclusive.

- Diesel, and other heavy oil derivatives, are known to be informally used at times, but are usually banned for this purpose.

References

- Bleasdale, J. K. A.; Salter, Peter John (1 January 1991). The Complete Know and Grow Vegetables. Oxford University Press. ISBN 978-0-19-286114-6.

- Ross, Merrill A.; Lembi, Carole A. (2008). Applied Weed Science: Including the Ecology and Management of Invasive Plants. Prentice Hall. p. 123. ISBN 978-0135028148.

- Quastel, J. H. (1950). "2,4-Dichlorophenoxyacetic Acid (2,4-D) as a Selective Herbicide". Agricultural Control Chemicals. Advances in Chemistry. 1. p. 244. doi:10.1021/ba-1950-0001.ch045. ISBN 0-8412-2442-0.

- Shaner, D. L.; Leonard, P. (2001). "Regulatory aspects of resistance management for herbicides and other crop protection products". In Powles, S. B.; Shaner, D. L. Herbicide Resistance and World Grains. CRC Press, Boca Raton, FL. pp. 279–294. ISBN 9781420039085.

- Powles, S. B.; Shaner, D. L., eds. (2001). Herbicide Resistance and World Grains. CRC Press, Boca Raton, FL. p. 328. ISBN 9781420039085.

- Moss, S. R. (2002). "Herbicide-Resistant Weeds". In Naylor,, R. E. L. Weed management handbook (9th ed.). Blackwell Science Ltd. pp. 225–252. ISBN 0-632-05732-7.

- "CDC - Pesticide Illness & Injury Surveillance - NIOSH Workplace Safety and Health Topic". Cdc.gov. 2013-09-11. Retrieved 2014-01-28.

- "Pesticides 101 - A primer on pesticides, their use in agriculture and the exposure we face | Pesticide Action Network". Panna.org. Retrieved 2014-01-28.

- Pal, GK; Kumar, B (2013). "Antifungal activity of some common weed extracts against wilt causing fungi, Fusarium oxysporum" (PDF). Current Discovery. International Young Scientist Association for Applied Research and Development. 2 (1): 62–67. ISSN 2320-4400. Retrieved February 8, 2014.

New Frontiers in Permaculture

An alternative form of farming, permaculture aims to provide solutions to the problem of over-exploitation of the land as well as the use of insecticides. Permaculture seeks to work alongside nature to design solutions for better cultivation and to also re-invest into land any surplus that is gained from agricultural practice. This chapter will provide an integrated understanding of permaculture.

Permaculture is a system of agricultural and social design principles centered on simulating or directly utilizing the patterns and features observed in natural ecosystems. The term *permaculture* (as a systematic method) was first coined by David Holmgren, then a graduate student, and his professor, Bill Mollison, in 1978. The word *permaculture* originally referred to "permanent agriculture", but was expanded to stand also for "permanent culture", as it was understood that social aspects were integral to a truly sustainable system as inspired by Masanobu Fukuoka's natural farming philosophy.

It has many branches that include but are not limited to ecological design, ecological engineering, environmental design, construction and integrated water resources management that develops sustainable architecture, regenerative and self-maintained habitat and agricultural systems modeled from natural ecosystems.

Mollison has said: "Permaculture is a philosophy of working with, rather than against nature; of protracted and thoughtful observation rather than protracted and thoughtless labor; and of looking at plants and animals in all their functions, rather than treating any area as a single product system."

History

In 1929, Joseph Russell Smith took up an antecedent term as the subtitle for *Tree Crops: A Permanent Agriculture*, a book in which he summed up his long experience experimenting with fruits and nuts as crops for human food and animal feed. Smith saw the world as an inter-related whole and suggested mixed systems of trees and crops underneath. This book inspired many individuals intent on making agriculture more sustainable, such as Toyohiko Kagawa who pioneered forest farming in Japan in the 1930s.

The definition of permanent agriculture as that which can be sustained indefinitely was supported by Australian P. A. Yeomans in his 1964 book *Water for Every Farm*. Yeomans introduced an observation-based approach to land use in Australia in the 1940s, and the keyline design as a way of managing the supply and distribution of water in the 1950s.

Stewart Brand's works were an early influence noted by Holmgren. Other early influences include Ruth Stout and Esther Deans, who pioneered no-dig gardening, and Masanobu Fukuoka who, in the late 1930s in Japan, began advocating no-till orchards, gardens and natural farming.

Core Tenets and Principles of Design

The three core tenets of permaculture are:

- Care for the earth: Provision for all life systems to continue and multiply. This is the first principle, because without a healthy earth, humans cannot flourish.

- Care for the people: Provision for people to access those resources necessary for their existence.

- Return of surplus: Reinvesting surpluses back into the system to provide for the first two ethics. This includes returning waste back into the system to recycle into usefulness. The third ethic is sometimes referred to as Fair Share to reflect that each of us should take no more than what we need before we reinvest the surplus.

Permaculture design emphasizes patterns of landscape, function, and species assemblies. It determines where these elements should be placed so they can provide maximum benefit to the local environment. The central concept of permaculture is maximizing useful connections between components and synergy of the final design. The focus of permaculture, therefore, is not on each separate element, but rather on the relationships created among elements by the way they are placed together; the whole becoming greater than the sum of its parts. Permaculture design therefore seeks to minimize waste, human labor, and energy input by building systems with maximal benefits between design elements to achieve a high level of synergy. Permaculture designs evolve over time by taking into account these relationships and elements and can become extremely complex systems that produce a high density of food and materials with minimal input.

The design principles which are the conceptual foundation of permaculture were derived from the science of systems ecology and study of pre-industrial examples of sustainable land use. Permaculture draws from several disciplines including organic farming, agroforestry, integrated farming, sustainable development, and applied ecology. Permaculture has been applied most commonly to the design of housing and landscaping, integrating techniques such as agroforestry, natural building, and rainwater harvesting within the context of permaculture design principles and theory.

Theory

Twelve Design Principles

Twelve Permaculture design principles articulated by David Holmgren in his *Permaculture: Principles and Pathways Beyond Sustainability*:

1. *Observe and interact*: By taking time to engage with nature we can design solutions that suit our particular situation.

2. *Catch and store energy*: By developing systems that collect resources at peak abundance, we can use them in times of need.

3. *Obtain a yield*: Ensure that you are getting truly useful rewards as part of the work that you are doing.

4. *Apply self-regulation and accept feedback*: We need to discourage inappropriate activity to ensure that systems can continue to function well.

5. *Use and value renewable resources and services*: Make the best use of nature's abundance to reduce our consumptive behavior and dependence on non-renewable resources.

6. *Produce no waste*: By valuing and making use of all the resources that are available to us, nothing goes to waste.

7. *Design from patterns to details*: By stepping back, we can observe patterns in nature and society. These can form the backbone of our designs, with the details filled in as we go.

8. *Integrate rather than segregate*: By putting the right things in the right place, relationships develop between those things and they work together to support each other.

9. *Use small and slow solutions*: Small and slow systems are easier to maintain than big ones, making better use of local resources and producing more sustainable outcomes.

10. *Use and value diversity*: Diversity reduces vulnerability to a variety of threats and takes advantage of the unique nature of the environment in which it resides.

11. *Use edges and value the marginal*: The interface between things is where the most interesting events take place. These are often the most valuable, diverse and productive elements in the system.

12. *Creatively use and respond to change*: We can have a positive impact on inevitable change by carefully observing, and then intervening at the right time.

Layers

Suburban permaculture garden in Sheffield, UK with different layers of vegetation

Layers are one of the tools used to design functional ecosystems that are both sustainable and of direct benefit to humans. A mature ecosystem has a huge number of relationships between its component parts: trees, understory, ground cover, soil, fungi, insects, and animals. Because plants grow to different heights, a diverse community of life is able to grow in a relatively small space, as the vegetation occupies different layers. There are generally seven recognized layers in a food forest, although some practitioners also include fungi as an eighth layer.

1. The canopy: the tallest trees in the system. Large trees dominate but typically do not saturate the area, i.e. there exist patches barren of trees.

2. Understory layer: trees that revel in the dappled light under the canopy.

3. Shrub layer: a diverse layer of woody perennials of limited height. includes most berry bushes.

4. Herbaceous layer: Plants in this layer die back to the ground every winter (if winters are cold enough, that is). They do not produce woody stems as the Shrub layer does. Many culinary and medicinal herbs are in this layer. A large variety of beneficial plants fall into this layer. May be annuals, biennials or perennials.

5. Soil surface/Groundcover: There is some overlap with the Herbaceous layer and the Groundcover layer; however plants in this layer grow much closer to the ground, grow densely to fill bare patches of soil, and often can tolerate some foot traffic. Cover crops retain soil and lessen erosion, along with green manures that add nutrients and organic matter to the soil, especially nitrogen.

6. Rhizosphere: Root layers within the soil. The major components of this layer are the soil and the organisms that live within it such as plant roots (including root crops such as potatoes and other edible tubers), fungi, insects, nematodes, worms, etc.

7. Vertical layer: climbers or vines, such as runner beans and lima beans (vine varieties).

Guilds

There are many forms of guilds, including guilds of plants with similar functions (that could interchange within an ecosystem), but the most common perception is that of a mutual support guild. Such a guild is a group of species where each provides a unique set of diverse functions that work in conjunction, or harmony. Mutual support guilds are groups of plants, animals, insects, etc. that work well together. Some plants may be grown for food production, some have tap roots that draw nutrients up from deep in the soil, some are nitrogen-fixing legumes, some attract beneficial insects, and others repel harmful insects. When grouped together in a mutually beneficial arrangement, these plants form a guild. See Dave Jacke's work on edible forest gardens for more information on other guilds, specifically resource-partitioning and community-function guilds.

Edge Effect

The edge effect in ecology is the effect of the juxtaposition or placing side by side of contrasting environments on an ecosystem. Permaculturists argue that, where vastly differing systems meet, there is an intense area of productivity and useful connections. An example of this is the coast; where the land and the sea meet there is a particularly rich area that meets a disproportionate percentage of human and animal needs. So this idea is played out in permacultural designs by using spirals in the herb garden or creating ponds that have wavy undulating shorelines rather than a simple circle or oval (thereby increasing the amount of edge for a given area).

Zones

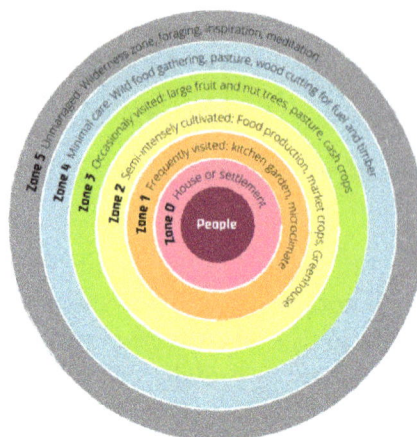

Permaculture Zones 0-5.

Zones are a way of intelligently organizing design elements in a human environment on the basis of the frequency of human use and plant or animal needs. Frequently manipulated or harvested elements of the design are located close to the house in zones 1 and 2. Less frequently used or manipulated elements, and elements that benefit from isolation (such as wild species) are farther away. Zones are about positioning things appropriately, and are numbered from 0 to 5.

Zone 0

> The house, or home center. Here permaculture principles would be applied in terms of aiming to reduce energy and water needs, harnessing natural resources such as sunlight, and generally creating a harmonious, sustainable environment in which to live and work. Zone 0 is an informal designation, which is not specifically defined in Bill Mollison's book.

Zone 1

> The zone nearest to the house, the location for those elements in the system that require frequent attention, or that need to be visited often, such as salad crops, herb plants, soft fruit like strawberries or raspberries, greenhouse and cold frames, propagation area, worm compost bin for kitchen waste, etc. Raised beds are often used in zone 1 in urban areas.

Zone 2

> This area is used for siting perennial plants that require less frequent maintenance, such as occasional weed control or pruning, including currant bushes and orchards, pumpkins, sweet potato, etc. This would also be a good place for beehives, larger scale composting bins, and so on.

Zone 3

> The area where main-crops are grown, both for domestic use and for trade purposes. After establishment, care and maintenance required are fairly minimal (provided mulches and similar things are used), such as watering or weed control maybe once a week.

Zone 4

A semi-wild area. This zone is mainly used for forage and collecting wild food as well as production of timber for construction or firewood.

Zone 5

A wilderness area. There is no human intervention in zone 5 apart from the observation of natural ecosystems and cycles. Through this zone we build up a natural reserve of bacteria, moulds and insects that can aid the zones above it.

People and Permaculture

Permaculture uses observation of nature to create regenerative systems, and the place where this has been most visible has been on the landscape. There has been a growing awareness though that firstly, there is the need to pay more attention to the peoplecare ethic, as it is often the dynamics of people that can interfere with projects, and secondly that the principles of permaculture can be used as effectively to create vibrant, healthy and productive people and communities as they have been in landscapes.

Domesticated Animals

Domesticated animals are often incorporated into site design.

Common Practices

Agroforestry

Agroforestry is an integrated approach of using the interactive benefits from combining trees and shrubs with crops and/or livestock. It combines agricultural and forestry technologies to create more diverse, productive, profitable, healthy and sustainable land-use systems. In agroforestry systems, trees or shrubs are intentionally used within agricultural systems, or non-timber forest products are cultured in forest settings.

Forest gardening is a term permaculturalists use to describe systems designed to mimic natural forests. Forest gardens, like other permaculture designs, incorporate processes and relationships that the designers understand to be valuable in natural ecosystems. The terms forest garden and food forest are used interchangeably in the permaculture literature. Numerous permaculturists are proponents of forest gardens, such as Graham Bell, Patrick Whitefield, Dave Jacke, Eric Toensmeier and Geoff Lawton. Bell started building his forest garden in 1991 and wrote the book *The Permaculture Garden* in 1995, Whitefield wrote the book *How to Make a Forest Garden* in 2002, Jacke and Toensmeier co-authored the two volume book set *Edible Forest Gardening* in 2005, and Lawton presented the film *Establishing a Food Forest* in 2008.

Tree Gardens, such as Kandyan tree gardens, in South and Southeast Asia, are often hundreds of years old. Whether they derived initially from experiences of cultivation and forestry, as is the case in agroforestry, or whether they derived from an understanding of forest ecosystems, as is the case for permaculture systems, is not self-evident. Many studies of these systems, especially those that predate the term permaculture, consider these systems to be forms of agroforestry.

Permaculturalists who include existing and ancient systems of polycropping with woody species as examples of food forests may obscure the distinction between permaculture and agroforestry.

Food forests and agroforestry are parallel approaches that sometimes lead to similar designs.

Hügelkultur

Hügelkultur is the practice of burying large volumes of wood to increase soil water retention. The porous structure of wood acts as a sponge when decomposing underground. During the rainy season, masses of buried wood can absorb enough water to sustain crops through the dry season. This technique has been used by permaculturalists Sepp Holzer, Toby Hemenway, Paul Wheaton, and Masanobu Fukuoka.

Natural Building

A natural building involves a range of building systems and materials that place major emphasis on sustainability. Ways of achieving sustainability through natural building focus on durability and the use of minimally processed, plentiful or renewable resources, as well as those that, while recycled or salvaged, produce healthy living environments and maintain indoor air quality.

The basis of natural building is the need to lessen the environmental impact of buildings and other supporting systems, without sacrificing comfort, health or aesthetics. To be more sustainable, natural building uses primarily abundantly available, renewable, reused or recycled materials. In addition to relying on natural building materials, the emphasis on the architectural design is heightened. The orientation of a building, the utilization of local climate and site conditions, the emphasis on natural ventilation through design, fundamentally lessen operational costs and positively impact the environment. Building compactly and minimizing the ecological footprint is common, as are on-site handling of energy acquisition, on-site water capture, alternate sewage treatment and water reuse.

Rainwater Harvesting

Rainwater harvesting is the accumulating and storing of rainwater for reuse before it reaches the aquifer. It has been used to provide drinking water, water for livestock, water for irrigation, as well as other typical uses. Rainwater collected from the roofs of houses and local institutions can make an important contribution to the availability of drinking water. It can supplement the subsoil water level and increase urban greenery. Water collected from the ground, sometimes from areas which are especially prepared for this purpose, is called stormwater harvesting.

Greywater is wastewater generated from domestic activities such as laundry, dishwashing, and bathing, which can be recycled on-site for uses such as landscape irrigation and constructed wetlands. Greywater is largely sterile, but not potable (drinkable). Greywater differs from water from the toilets which is designated sewage or blackwater, to indicate it contains human waste. Blackwater is septic or otherwise toxic and cannot easily be reused. There are, however, continuing efforts to make use of blackwater or human waste. The most notable is for composting through a process known as humanure; a combination of the words human and manure. Additionally, the methane in humanure can be collected and used similar to natural gas as a fuel, such as for heating or cooking, and is commonly referred to as biogas. Biogas can be harvested from the human waste and the remainder still used as humanure. Some of the simplest forms of humanure use include a composting toilet or an outhouse or dry bog surrounded by trees that are heavy feeders which can be coppiced for wood fuel. This process eliminates the use of a standard toilet with plumbing.

Sheet Mulching

In agriculture and gardening, mulch is a protective cover placed over the soil. Any material or combination can be used as mulch, such as stones, leaves, cardboard, wood chips, gravel, etc., though in permaculture mulches of organic material are the most common because they perform more functions. These include: absorbing rainfall, reducing evaporation, providing nutrients, increasing organic matter in the soil, feeding and creating habitat for soil organisms, suppressing weed growth and seed germination, moderating diurnal temperature swings, protecting against frost, and reducing erosion. Sheet mulching is an agricultural no-dig gardening technique that attempts to mimic natural processes occurring within forests. Sheet mulching mimics the leaf cover that is found on forest floors. When deployed properly and in combination with other Permacultural principles, it can generate healthy, productive and low maintenance ecosystems.

Sheet mulch serves as a "nutrient bank," storing the nutrients contained in organic matter and slowly making these nutrients available to plants as the organic matter slowly and naturally breaks down. It also improves the soil by attracting and feeding earthworms, slaters and many other soil micro-organisms, as well as adding humus. Earthworms "till" the soil, and their worm castings are among the best fertilizers and soil conditioners. Sheet mulching can be used to reduce or eliminate undesirable plants by starving them of light, and can be more advantageous than using herbicide or other methods of control.

Intensive Rotational Grazing

Grazing has long been blamed for much of the destruction we see in the environment. However, it has been shown that when grazing is modeled after nature, the opposite effect can be seen. Also known as cell grazing, managed intensive rotational grazing (MIRG) is a system of grazing in which ruminant and non-ruminant herds and/or flocks are regularly and systematically moved to fresh pasture, range, or forest with the intent to maximize the quality and quantity of forage growth. This disturbance is then followed by a period of rest which allows new growth. MIRG can be used with cattle, sheep, goats, pigs, chickens, rabbits, geese, turkeys, ducks and other animals depending on the natural ecological community that is being mimicked. Sepp Holzer and Joel Salatin have shown how the disturbance caused by the animals can be the spark needed to start ecological succession or prepare ground for planting. Allan Savory's holistic management technique has been likened to "a permaculture approach to rangeland management". One variation on MIRG that is gaining rapid popularity is called eco-grazing. Often used to either control invasives or re-establish native species, in eco-grazing the primary purpose of the animals is to benefit the environment and the animals can be, but are not necessarily, used for meat, milk or fiber.

Keyline Design

Keyline design is a technique for maximizing beneficial use of water resources of a piece of land developed in Australia by farmer and engineer P. A. Yeomans. The *Keyline* refers to a specific topographic feature linked to water flow which is used in designing the drainage system of the site.

Fruit Tree Management

The no-pruning option is usually ignored by fruit experts, though often practised by default in people's back gardens! But it has its advantages. Obviously it reduces work, and more surprisingly it can lead to higher overall yields.

— *Whitefield, Patrick, How to make a forest garden, p. 16*

Masanobu Fukuoka, as part of early experiments on his family farm in Japan, experimented with no-pruning methods, noting that he ended up killing many fruit trees by simply letting them go, which made them become convoluted and tangled, and thus unhealthy. Then he realised this is the difference between natural-form fruit trees and the process of change of tree form that results from abandoning previously-pruned unnatural fruit trees. He concluded that the trees should be raised all their lives without pruning, so they form healthy and efficient branch patterns that follow their natural inclination. This is part of his implementation of the Tao-philosophy of Wú wéi translated in part as no-action (against nature), and he described it as no unnecessary pruning, nature farming or "do-nothing" farming, of fruit trees, distinct from non-intervention or literal no-pruning. He ultimately achieved yields comparable to or exceeding standard/intensive practices of using pruning and chemical fertilisation.

Another proponent of the no, or limited, pruning method is Sepp Holzer who used the method in connection with Hügelkultur berms. He has successfully grown several varieties of fruiting trees at altitudes (approximately 9,000 feet (2,700 m)) far above their normal altitude, temperature, and snow load ranges. He notes that the Hügelkultur berms kept and/or generated enough heat to allow the roots to survive during alpine winter conditions. The point of having unpruned branches, he notes, was that the longer (more naturally formed) branches bend over under the snow load until they touched the ground, thus forming a natural arch against snow loads that would break a shorter, pruned, branch.

Compost Management

Compost Basket

Compost Basket is a way to the permanent management of compost materials. The idea of Compost Basket comes from Gyulai Iván. He invented and used it in the Gömörszőlős Educational Center. The inner circle is 1 meter deep. This is the place where compost material are put in. Under this is the outer circle, 40 cm deep. Here the nutrients come down.

Mollison and Holmgren

Bill Mollison in January 2008.

In the mid-1970s, Bill Mollison and David Holmgren started developing ideas about stable agricultural systems on the southern Australian island state of Tasmania. This was a result of the danger of the rapidly growing use of industrial-agricultural methods. In their view, highly dependent on non renewable resources, these methods were additionally poisoning land and water, reducing biodiversity, and removing billions of tons of topsoil from previously fertile landscapes. A design approach called *permaculture* was their response and was first made public with the publication of their book *Permaculture One* in 1978.

By the early 1980s, the concept had broadened from agricultural systems design towards sustainable human habitats. After *Permaculture One*, Mollison further refined and developed the ideas by designing hundreds of permaculture sites and writing more detailed books, notably *Permaculture: A Designers Manual*. Mollison lectured in over 80 countries and taught his two-week Permaculture Design Course (PDC) to many hundreds of students.Mollison "encouraged graduates to become teachers themselves and set up their own institutes and demonstration sites. This multiplier effect was critical to permaculture's rapid expansion."

In 1991, a four-part television documentary by ABC productions called "The Global Gardener" showed permaculture applied to a range of worldwide situations, bringing the concept to a much broader public. In 2012, the UMass Permaculture Initiative won the White House "Champions of Change" sustainability contest, which declared that "they demonstrate how permaculture can feed a growing population in an environmentally sustainable and socially responsible manner".

In 1997, Holmgren explained that the primary agenda of the permaculture movement is to assist people to become more self-reliant through the design and development of productive and sustainable gardens and farms.

In 2014, Holmgren endorsed and helped launch a new Australian permaculture magazine, Pip Magazine.

Notable Permaculturists

Joseph Russell Smith took up an antecedent term as the subtitle for *Tree Crops: A Permanent Agriculture*, a book in which he summed up his long experience experimenting with fruits and nuts as crops for human food and animal feed. By that year (1929), Smith saw the world as an interrelated whole and suggested mixed systems of trees and crops underneath. This book inspired many individuals intent on making permaculture a valid means of sustainable food production. Bill Mollison and David Holmgren developed it further, and permaculturists were trained under the umbrella of Bill Mollison's train the trainer system.

Geoff Lawton, Toby Hemenway and P. A. Yeomans - creator of the keyline design each have more than 20 years experience teaching and promo ting permaculture as a sustainable way of growing food. Simon Fjell was a Founding Director of the Permaculture Institute in late 1979, over 40 years experience, having first met Mollison in 1976. He has since worked in every continent.

The permaculture movement also spread throughout Asia and Central America, with Hong Kong-based Asian Institute of Sustainable Architecture (AISA), Rony Lec leading the foundation of the Mesoamerican Permaculture Institute (IMAP) in Guatemala and Juan Rojas co-founding the Permaculture Institute of El Salvador.

Trademark and Copyright Issues

There has been contention over who, if anyone, controls legal rights to the word *permaculture*: is it trademarked or copyrighted? and if so, who holds the legal rights to the use of the word? For a long time Bill Mollison claimed to have copyrighted the word, and his books said on the copyright page, "The contents of this book and the word PERMACULTURE are copyright." These statements were largely accepted at face-value within the permaculture community. However, copyright law does not protect names, ideas, concepts, systems, or methods of doing something; it only protects the expression or the description of an idea, not the idea itself. Eventually Mollison acknowledged that he was mistaken and that no copyright protection existed for the word *permaculture*.

In 2000, Mollison's US based Permaculture Institute sought a service mark (a form of trademark) for the word *permaculture* when used in educational services such as conducting classes, seminars, or workshops. The service mark would have allowed Mollison and his two Permaculture Institutes (one in the US and one in Australia) to set enforceable guidelines regarding how permaculture could be taught and who could teach it, particularly with relation to the PDC, despite the fact that he had instituted a system of certification of teachers to teach the PDC in 1993. The service mark failed and was abandoned in 2001. Also in 2001 Mollison applied for trademarks in Australia for the terms "Permaculture Design Course" and "Permaculture Design". These applications were both withdrawn in 2003. In 2009 he sought a trademark for "Permaculture: A Designers' Manual" and "Introduction to Permaculture", the names of two of his books. These applications were withdrawn in 2011. There has never been a trademark for the word *permaculture* in Australia.

Criticisms

General Criticisms

In 2011, Owen Hablutzel argued that "permaculture has yet to gain a large amount of specific mainstream scientific acceptance," and that "the sensitiveness to being perceived and accepted on scientific terms is motivated in part by a desire for permaculture to expand and become increasingly relevant." Bec-Hellouin permaculture farm engaged in a research program in partnership with INRA and AgroParisTech to collect scientific data.

In his books *Sustainable Freshwater Aquaculture* and *Farming in Ponds and Dams*, Nick Romanowski expresses the view that the presentation of aquaculture in Bill Mollison's books is unrealistic and misleading.

Agroforestry

Greg Williams argues that forests cannot be more productive than farmland because the net productivity of forests decline as they mature due to ecological succession. Proponents of permaculture respond that this is true only if one compares data between woodland forest and climax vegetation, but not when comparing farmland vegetation with woodland forest. For example, ecological succession generally results in a forest's productivity rising after its establishment only until it reaches the *woodland state* (67% tree cover), before declining until *full maturity*.

References

- Holmgren, David (2002). Permaculture: Principles & Pathways Beyond Sustainability. Holmgren Design Services. p. 1. ISBN 0-646-41844-0.

- Ash, Andrew, The Ecograze Project – developing guidelines to better manage grazing country (PDF), et al., CSIRO, ISBN 0-9579842-0-0, retrieved 7 April 2013

- Nick Romanowski (2007). Sustainable Freshwater Aquaculture: The Complete Guide from Backyard to Investor. UNSW Press. p. 130. ISBN 978-0-86840-835-4.

- Lillington, Ian; Holmgren, David; Francis, Robyn; Rosenfeldt, Robyn. "The Permaculture Story: From 'Rugged Individuals' to a Million Member Movement" (PDF). Pip Magazine. Retrieved 9 July 2015.

- "Prince Charles sends a message to IUCN's World Conservation Congress". International Union for Conservation of Nature. Retrieved 6 April 2013.

- Undersander, Dan; et al. "Grassland birds: Fostering habitat using rotational grazing" (PDF). University of Wisconsin-Extension. Retrieved 5 April 2013.

- Gordon, Ian. "A systems approach to livestock/resource interactions in tropical pasture systems" (PDF). The James Hutton Institute. Retrieved 7 April 2013.

- Paul, Willi (2011). "Symbols & Patterns. Interview with Owen Hablutzel, Director, Permaculture Research Institute, USA". Retrieved 2012-06-21.

- Grayson, Russ (2011). "The Permaculture Papers 5: time of change and challenge — 2000-2004". Pacific edge. Retrieved 8 September 2011.

- United States Patent and Trademark Office (2011). "Trademark Electronic Search System (TESS)". US Department of Commerce. Retrieved 8 September 2011.

Harvest and Harvesting Technologies

Harvesting is a labour-intensive process and requires skill, acumen and effort. It is only natural then that the harvest is one of the stages that would see high amounts of mechanisation and automation of its various processes. Some of the themes discussed in this chapter are on reapers, threshing methods, combine harvesters and the practice of winnowing.

Harvest

Harvesting agriculture in Volgograd Oblast, Russia

Harvesting is the process of gathering a ripe crop from the fields. *Reaping* is the cutting of grain or pulse for harvest, typically using a scythe, sickle, or reaper. On smaller farms with minimal mechanization, harvesting is the most labor-intensive activity of the growing season. On large mechanized farms, harvesting utilizes the most expensive and sophisticated farm machinery, such as the combine harvester. The term "harvesting" in general usage may include immediate postharvest handling, including cleaning, sorting, packing, and cooling.

The completion of harvesting marks the end of the growing season, or the growing cycle for a particular crop, and the social importance of this event makes it the focus of seasonal celebrations such as harvest festivals, found in many religions.

Rye harvest on Gotland, Sweden, 1900–1910.

Sugar beet harvester. Baden-Wurttemberg, Germany.

Etymology

"Harvest", a noun, came from the Old English word *hærfest*, meaning "autumn" (the season), "harvest-time", or "August". (It continues to mean "autumn" in British dialect, and "season of gathering crops" generally.) "The harvest" came to also mean the activity of reaping, gathering, and storing grain and other grown products during the autumn, and also the grain and other grown products themselves. "Harvest" was also verbified: "To harvest" means to reap, gather, and store the harvest (or the crop). People who harvest and equipment that harvests are harvesters; while they do it, they are harvesting.

Crop Failure

Crop failure (also known as harvest failure) is an absent or greatly diminished crop yield relative to expectation, caused by the plants being damaged, killed, or destroyed, or affected in some way that they fail to form edible fruit, seeds, or leaves in their expected abundance.

Crop failures can be caused by catastrophic events such as plant disease outbreaks (e.g. Great Famine (Ireland)), heavy rainfall, volcanic eruptions, storms, floods, or drought, or by slow, cumulative effects of soil degradation, too-high soil salinity, erosion, desertification, usually as results of drainage, overdrafting (for irrigation), overfertilization, or overexploitation.

In history, crop failures and subsequent famines have triggered human migration, rural exodus, etc.

The proliferation of industrial monocultures, with their reduction in crop diversity and dependence on heavy use of artificial fertilizers and pesticides, has led to overexploited soils that are nearly incapable of regeneration. Over years, unsustainable farming of land degrades soil fertility and diminishes crop yield. With a steadily growing world population and local overpopulation, even slightly diminishing yields are already the equivalent to a partial harvest failure.

Other Uses

Harvesting commonly refers to grain and produce, but also has other uses. Fishing and logging are also referred to as harvesting. The term harvest is also used in reference to harvesting grapes for wine. Within the context of irrigation, *water harvesting* refers to the collection and run-off of rainwater for agricultural or domestic uses. Instead of *harvest*, the term *exploit* is also used, as in exploiting fisheries or water resources. *Energy harvesting* is the process of capturing and storing energy (such as solar power, thermal energy, wind energy, salinity gradients, and kinetic energy) that would otherwise go unexploited. *Body harvesting*, or *cadaver harvesting*, is the process of collecting and preparing cadavers for anatomical study. In a similar sense, *organ harvesting* is the removal of tissues or organs from a donor for purposes of transplanting.

Harvesting or *Domestic Harvesting* in Canada refers to hunting, fishing, and plant gathering by First Nations, Métis, and Inuit in discussions of aboriginal or treaty rights. For example, in the Gwich'in Comprehensive Land Claim Agreement, "Harvesting means gathering, hunting, trapping or fishing..." Similarly, in the Tlicho Land Claim and Self Government Agreement, "'Harvesting' means, in relation to wildlife, hunting, trapping or fishing and, in relation to plants or trees, gathering or cutting."

Growing Degree-day

Growing degree days (GDD), also called growing degree units (GDUs), are a heuristic tool in phenology. GDD are a measure of heat accumulation used by horticulturists, gardeners, and farmers to predict plant and animal development rates such as the date that a flower will bloom, an insect will emerge from dormancy, or a crop will reach maturity.

Introduction

In the absence of extreme conditions such as unseasonal drought or disease, plants grow in a cumulative stepwise manner which is strongly influenced by the ambient temperature. Growing degree days take aspects of local weather into account and allow gardeners to predict (or, in greenhouses, even to control) the plants' pace toward maturity.

Unless stressed by other environmental factors like moisture, the development rate from emergence to maturity for many plants depends upon the daily air temperature. Because many developmental events of plants and insects depend on the accumulation of specific quantities of heat, it is possible to predict when these events should occur during a growing season regardless of differences in temperatures from year to year. Growing degrees (GDs) is defined as the number of temperature degrees above a certain threshold base temperature, which varies among crop species. The base temperature is that temperature below which plant growth is zero. GDs are calculated each day as maximum temperature plus the minimum temperature divided by 2 (or the mean temperature), minus the base temperature. GDUs are accumulated by adding each day's GDs contribution as the season progresses.

GDUs can be used to: assess the suitability of a region for production of a particular crop; estimate the growth-stages of crops, weeds or even life stages of insects; predict maturity and cutting dates of forage crops; predict best timing of fertilizer or pesticide application; estimate the heat stress on crops; plan spacing of planting dates to produce separate harvest dates. Crop specific indices that employ separate equations for the influence of the daily minimum (nighttime) and the maximum (daytime) temperatures on growth are called crop heat units (CHUs).

GDD Calculation

GDD are calculated by taking the integral of warmth above a base temperature, T_{base} (usually 10 °C):

$$GDD = \int (T - T_{base})dt.$$

A simpler, approximately equivalent formulation uses the average of the daily maximum and minimum temperatures compared to a T_{base}. As an equation:

$$GDD = \frac{T_{max} + T_{min}}{2} - T_{base}.$$

If the mean daily temperature is lower than the base temperature then GDD=0.

GDDs are typically measured from the winter low. Any temperature below T_{base} is set to T_{base} before calculating the average. Likewise, the maximum temperature is usually capped at 30 °C because most plants and insects do not grow any faster above that temperature. However, some warm temperate and tropical plants do have significant requirements for days above 30 °C to mature fruit or seeds.

For example, a day with a high of 23 °C and a low of 12 °C (and a base of 10 °C) would contribute 7.5 GDDs.

$$\frac{23+12}{2} - 10 = 7.5$$

A day with a high of 13 °C and a low of 10 °C (and a base of 10 °C) would contribute 1.5 GDDs.

$$\frac{13+10}{2} - 10 = 1.5$$

Plant development

Common name	Latin name	Number of growing degree days baseline 10 °C
Witch-hazel	*Hamamelis* spp.	begins flowering at <1 GDD
Red maple	*Acer rubrum*	begins flowering at 1-27 GDD
Forsythia	*Forsythia* spp.	begin flowering at 1-27 GDD
Sugar maple	*Acer saccharum*	begin flowering at 1-27 GDD
Norway maple	*Acer platanoides*	begins flowering at 30-50 GDD
White ash	*Fraxinus americana*	begins flowering at 30-50 GDD
Crabapple	*Malus* spp.	begins flowering at 50-80 GDD
Common Broom	*Cytisus scoparius*	begins flowering at 50-80 GDD
Horsechestnut	*Aesculus hippocastanum*	begin flowering at 80-110 GDD
Common lilac	*Syringa vulgaris*	begin flowering at 80-110 GDD
Beach plum	*Prunus maritima*	full bloom at 80-110 GDD
Black locust	*Robinia pseudoacacia*	begins flowering at 140-160 GDD
Catalpa	*Catalpa speciosa*	begins flowering at 250-330 GDD
Privet	*Ligustrum* spp.	begins flowering at 330-400 GDD
Elderberry	*Sambucus canadensis*	begins flowering at 330-400 GDD
Purple loosestrife	*Lythrum salicaria*	begins flowering at 400-450 GDD
Sumac	*Rhus typhina*	begins flowering at 450-500 GDD
Butterfly bush	*Buddleia davidii*	begins flowering at 550-650 GDD
Corn (maize)	*Zea mays*	800 to 1400 GDD to crop maturity
Dry beans	*Phaseolus vulgaris*	1100-1300 GDD to maturity depending on cultivar and soil conditions
Sugar Beet	*Beta vulgaris*	130 GDD to emergence and 1400-1500 GDD to maturity
Barley	*Hordeum vulgare*	125-162 GDD to emergence and 1290-1540 GDD to maturity
Wheat (Hard Red)	*Triticum aestivum*	143-178 GDD to emergence and 1550-1680 GDD to maturity (there may be confusion between Celsius and Fahrenheit, 1680 is probably Fahrenheit)
Oats	*Avena sativa*	1500-1750 GDD to maturity
European Corn Borer	*Ostrinia nubilalis*	207 - Emergence of first spring moths

Pest Control

- Insect development and growing degree days are also used by some farmers and horticulturalists to time their use of organic or biological pest control or other pest control methods so they are applying the procedure or treatment at the point that the pest is most vulnerable. For example:

- Black cutworm larvae have grown large enough to start causing economic damage at 165 GDD

- Azalea Lace Bug emerges at about 130 GDD

- Boxwood leaf miner emerges at about 250 GDD

Honeybees

Several beekeepers are now researching the correlation between growing degree days-GDD and the lifecycle of a honeybee colony.

Baselines

10 °C is the most common base for GDD calculations, however, the optimal base is often determined experimentally based on the lifecycle of the plant or insect in question.

- 5.5 °C wheat, barley, rye, oats, flaxseed, lettuce, asparagus

- 6 °C Stalk Borer

- 7 °C Corn rootworm

- 8 °C sunflower, potato

- 9 °C Alfalfa weevil

- 10 °C maize (including sweet corn), sorghum, rice, soybeans, tomato, Black cutworm, European Corn Borer, Coffee (Jaramillo-Robledo & Guzman-Martinez published by Cenicafé), standard baseline for insect and mite pests of woody plants

- 11 °C Green Cloverworm

- 12 °C many other crop calculations

- 30 °C the USDA measure heat zones in GDD above 30 °C; for many plants this is significant for seed maturation, e.g. reed (*Phragmites*) requires at least some days reaching this temperature to mature viable seeds

GDDs may be calculated using either Celsius or Fahrenheit, though they must be converted appropriately; for every 9 GDD^F there is 5 GDD^C, or in conversion calculation:

$$GDD^C = 5/9 * GDD^F$$

Combine Harvester

The combine harvester, or simply combine, is a machine that harvests grain crops. The name derives from its combining three separate operations comprising harvesting—reaping, threshing, and winnowing—into a single process. Among the crops harvested with a combine are wheat, oats, rye, barley, corn (maize), sorghum, soybeans, flax (linseed), sunflowers, and canola. The waste straw left behind on the field is the remaining dried stems and leaves of the crop with limited nutrients which is either chopped and spread on the field or baled for feed and bedding for livestock.

A Lely open-cab combine.

Harvesting oats with a Claas Lexion 570 harvester with enclosed air-conditioned cab, rotary thresher and laser-guided automatic steering

Old Style Harvester found in the Henty, Australia region

John Deere Combine 9870 STS with 625D

Rostselmash Combine Torum 740

John Deere 9870 STS underbelly

Case IH Axial-Flow combine

Combine harvesters are one of the most economically important labour saving inventions, significantly reducing the fraction of the population that must be engaged in agriculture.

History

Scottish inventor Patrick Bell invented the reaper in 1826. The combine was invented in the United States by Hiram Moore in 1834. Early versions were pulled by horse, mule or ox teams. In 1835, Moore built a full-scale version and by 1839, over 20 ha (50 acres) of crops were harvested. By 1860, combine harvesters with a cutting, or swathe, width of several metres were used on American farms. Australian Hugh Victor McKay produced a commercially successful combine harvester in 1885, the Sunshine Harvester.

Combines, some of them quite large, were drawn by mule or horse teams and used a bullwheel to provide power. Later, steam power was used, and George Stockton Berry integrated the combine with a steam engine using straw to heat the boiler. At the turn of the twentieth century, horse drawn combines were starting to be used on the American plains and Idaho (often pulled by teams of twenty or more horses).

In 1911, the Holt Manufacturing Company of California produced a self-propelled harvester. In Australia in 1923, the patented Sunshine Auto Header was one of the first center-feeding self-propelled harvesters. In 1923 in Kansas, the Baldwin brothers and their Gleaner Manufacturing Company patented a self-propelled harvester that included several other modern improvements in grain handling. Both the Gleaner and the Sunshine used Fordson engines; early Gleaners used the entire Fordson chassis and driveline as a platform. In 1929 Alfredo Rotania of Argentina patented a self-propelled harvester. International Harvester started making horse-pulled combines in 1915. At the time horse powered binders and stand alone threshing machines were more common. In the 1920s Case Corporation and John Deere made combines and these were starting to be tractor pulled with a second engine aboard the combine to power its workings. The world economic collapse in the 1930s stopped farm equipment purchases thus people largely retained the older method of harvesting. A few farms did invest and used Caterpillar tractors to move the outfits.

Tractor-drawn combines (also called pull-type combines) became common after World War II as many farms began to use tractors. An example was the All-Crop Harvester series. These combines used a shaker to separate the grain from the chaff and straw-walkers (grates with small teeth on an eccentric shaft) to eject the straw while retaining the grain. Early tractor-drawn combines were usually powered by a separate gasoline engine, while later models were PTO-powered. These machines either put the harvested crop into bags that were then loaded onto a wagon or truck, or had a small bin that stored the grain until it was transferred to a truck or wagon with an auger.

In the U.S., Allis-Chalmers, Massey-Harris, International Harvester, Gleaner Manufacturing Company, John Deere, and Minneapolis Moline are past or present major combine producers. In 1937, the Australian-born Thomas Carroll, working for Massey-Harris in Canada, perfected a self-propelled model and in 1940 a lighter-weight model began to be marketed widely by the company. Lyle Yost invented an auger that would lift grain out of a combine in 1947, making unloading grain much easier. In 1952 Claeys launched the first self-propelled combine harvester in Europe; in 1953, the European manufacturer CLAAS developed a self-propelled combine harvester named 'Herkules', it could harvest up to 5 tons of wheat a day. This newer kind of combine is still in use and is powered by diesel or gasoline engines. Until the self-cleaning rotary screen was invented in the mid-1960s combine engines suffered from overheating as the chaff spewed out when harvesting small grains would clog radiators, blocking the airflow needed for cooling.

A significant advance in the design of combines was the rotary design. The grain is initially stripped from the stalk by passing along a helical rotor instead of passing between rasp bars on the outside of a cylinder and a concave. Rotary combines were first introduced by Sperry-New Holland in 1975.

In about the 1980s on-board electronics were introduced to measure threshing efficiency. This new instrumentation allowed operators to get better grain yields by optimizing ground speed and other operating parameters.

A New Holland TX68 with grain platform attached.

A John Deere Titan series combine unloading corn.

Combine Heads

Combines are equipped with removable heads that are designed for particular crops. The standard header, sometimes called a grain platform, is equipped with a *reciprocating knife cutter bar*, and features a revolving reel with metal or plastic teeth to cause the cut crop to fall into the auger once it is cut. A variation of the platform, a "flex" platform, is similar but has a cutter bar that can flex over contours and ridges to cut soybeans that have pods close to the ground. A flex head can cut soybeans as well as cereal crops, while a rigid platform is generally used only in cereal grains.

Some wheat headers, called "draper" headers, use a fabric or rubber apron instead of a cross auger. Draper headers allow faster feeding than cross augers, leading to higher throughputs due to lower power requirements. On many farms, platform headers are used to cut wheat, instead of separate wheat headers, so as to reduce overall costs.

Dummy heads or pick-up headers feature spring-tined pickups, usually attached to a heavy rubber belt. They are used for crops that have already been cut and placed in windrows or swaths. This is particularly useful in northern climates such as western Canada where swathing kills weeds resulting in a faster dry down.

While a grain platform can be used for corn, a specialized corn head is ordinarily used instead. The corn head is equipped with snap rolls that strip the stalk and leaf away from the ear, so that only the ear (and husk) enter the throat. This improves efficiency dramatically since so much less material must go through the cylinder. The corn head can be recognized by the presence of points between each row.

Occasionally rowcrop heads are seen that function like a grain platform, but have points between rows like a corn head. These are used to reduce the amount of weed seed picked up when harvesting small grains.

Self-propelled Gleaner combines could be fitted with special tracks instead of tires or tires with tread measuring almost 10in deep to assist in harvesting rice. Some combines, particularly pull type, have tires with a diamond tread which prevents sinking in mud. These tracks can fit other combines by having adapter plates made.

Conventional Combine

The cut crop is carried up the feeder throat (commonly called the "feederhouse") by a *chain and flight elevator*, then fed into the threshing mechanism of the combine, consisting of a rotating *threshing drum* (commonly called the "cylinder"), to which grooved steel bars (rasp bars) are bolted. The rasp bars thresh or separate the grains and chaff from the straw through the action of the cylinder against the *concave*, a shaped "half drum", also fitted with steel bars and a meshed grill, through which grain, chaff and smaller debris may fall, whereas the straw, being too long, is carried through onto the *straw walkers*. This action is also allowed due to the fact that the grain is heavier than the straw, which causes it to fall rather than "float" across from the cylinder/concave to the walkers. The drum speed is variably adjustable on most machines, whilst the distance between the drum and concave is finely adjustable fore, aft and together, to achieve optimum separation and output. Manually engaged *disawning plates* are usually fitted to the concave. These provide extra friction to remove the awns from barley crops. After the primary separation at the cylinder, the clean grain falls through the concave and to the shoe, which contains the chaffer and sieves. The shoe is common to both conventional combines and rotary combines.

Hillside Leveling

In the Palouse region of the Pacific Northwest of the United States the combine is retrofitted with a hydraulic hillside leveling system. This allows the combine to harvest the steep but fertile soil in the region. Hillsides can be as steep as a 50% slope. Gleaner, IH and Case IH, John Deere, and others all have made combines with this hillside leveling system, and local machine shops have fabricated them as an aftermarket add-on.

Palouse hills

A Massey Ferguson combine fitted with the hillside leveling option

The first leveling technology was developed by Holt Co., a California firm, in 1891. Modern leveling came into being with the invention and patent of a level sensitive mercury switch system invented by Raymond Alvah Hanson in 1946. Raymond's son, Raymond, Jr., produced leveling systems exclusively for John Deere combines until 1995 as R. A. Hanson Company, Inc. In 1995, his son, Richard, purchased the company from his father and renamed it RAHCO International, Inc. In March 2011, the company was renamed Hanson Worldwide, LLC. Production continues to this day.

Hillside leveling has several advantages. Primary among them is an increased threshing efficiency on hillsides. Without leveling, grain and chaff slide to one side of separator and come through the machine in a large ball rather than being separated, dumping large amounts of grain on the ground. By keeping the machinery level, the straw-walker is able to operate more efficiently, making for more efficient threshing. IH produced the 453 combine which leveled both side-to-side and front-to-back, enabling efficient threshing whether on a hillside or climbing a hill head on.

Secondarily, leveling changes a combine's center of gravity relative to the hill and allows the combine to harvest along the contour of a hill without tipping, a very real danger on the steeper slopes of the region; it is not uncommon for combines to roll on extremely steep hills.

Newer leveling systems do not have as much tilt as the older ones. A John Deere 9600 combine equipped with a Rahco hillside conversion kit will level over to 44%, while the newer STS combines will only go to 35%. These modern combines use the rotary grain separator which makes leveling less critical. Most combines on the Palouse have dual drive wheels on each side to stabilize them.

A leveling system was developed in Europe by the Italian combine manufacturer Laverda which still produces it today.

Sidehill Leveling

Sidehill combines are very similar to hillside combines in that they level the combine to the ground so that the threshing can be efficiently conducted; however, they have some very distinct differences. Modern hillside combines level around 35% on average, older machines were closer to 50%. Sidehill combines only level to 18%. They are sparsely used in the Palouse region. Rather, they are used on the gentle rolling slopes of the mid-west. Sidehill combines are much more mass-produced than their hillside counterparts. The height of a sidehill machine is the same height as a level-land combine. Hillside combines have added steel that sets them up approximately 2–5 feet higher than a level-land combine and provide a smooth ride

Maintaining Threshing Speed

Allis-Chalmers GLEANER L2

Another technology that is sometimes used on combines is a continuously variable transmission. This allows the ground speed of the machine to be varied while maintaining a constant engine and threshing speed. It is desirable to keep the threshing speed constant since the machine will typically have been adjusted to operate best at a certain speed.

Self-propelled combines started with standard manual transmissions that provided one speed based on input rpm. Deficiencies were noted and in the early 1950s combines were equipped with what John Deere called the "Variable Speed Drive". This was simply a variable width sheave controlled by spring and hydraulic pressures. This sheave was attached to the input shaft of the transmission. A standard 4 speed manual transmission was still used in this drive system. The operator would select a gear, typically 3rd. An extra control was provided to the operator to allow him to speed up and slow down the machine within the limits provided by the variable speed drive system.

By decreasing the width of the sheave on the input shaft of the transmission, the belt would ride higher in the groove. This slowed the rotating speed on the input shaft of the transmission, thus slowing the ground speed for that gear. A clutch was still provided to allow the operator to stop the machine and change transmission gears.

Later, as hydraulic technology improved, hydrostatic transmissions were introduced by Versatile Mfg for use on swathers but later this technology was applied to combines as well. This drive retained the 4 speed manual transmission as before, but this time used a system of hydraulic pumps and motors to drive the input shaft of the transmission. This system is called a Hydrostatic drive system. The engine turns the hydraulic pump capable of pressures up to 4,000 psi (30 MPa). This pressure is then directed to the hydraulic motor that is connected to the input shaft of the transmission. The operator is provided with a lever in the cab that allows for the control of the hydraulic motor's ability to use the energy provided by the pump. By adjusting the swash plate in the motor, the stroke of its pistons are changed. If the swash plate is set to neutral, the pistons do not move in their bores and no rotation is allowed, thus the machine does not move. By moving the lever, the swash plate moves its attached pistons forward, thus allowing them to move within the bore and causing the motor to turn. This provides an infinitely variable speed control from 0 ground speed to what ever the maximum speed is allowed by the gear selection of the transmission. The standard clutch was removed from this drive system as it was no longer needed.

Most if not all modern combines are equipped with hydrostatic drives. These are larger versions of the same system used in consumer and commercial lawn mowers that most are familiar with today. In fact, it was the downsizing of the combine drive system that placed these drive systems into mowers and other machines.

The Threshing Process

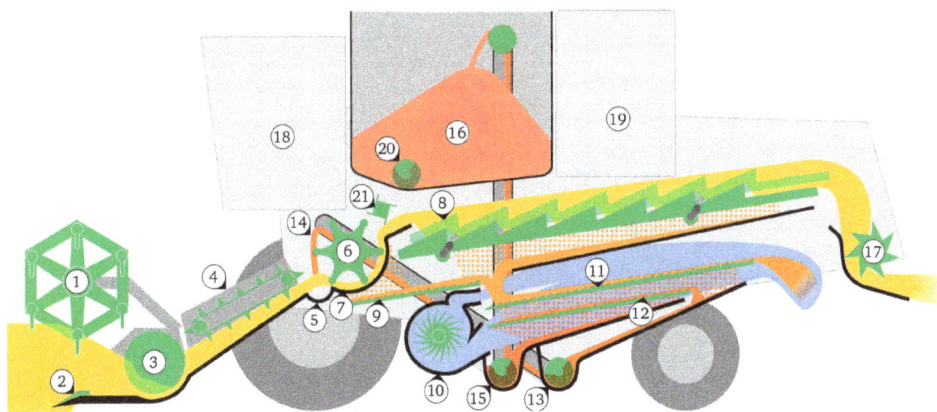

Conventional combine harvester (cut)

1) Reel	8) Straw walker	15) Grain auger
2) Cutter bar	9) Grain pan	16) Grain tank
3) Header auger	10) Fan	17) Straw chopper
4) Grain conveyor	11) Top Adjustable sieve	18) Driver's cab
5) Stone trap	12) Bottom sieve	19) Engine
6) Threshing drum	13) Tailings conveyor	20) Unloading auger
7) Concave	14) Rethreshing of tailings	21) Impeller

Despite great advances mechanically and in computer control, the basic operation of the combine harvester has remained unchanged almost since it was invented.

First, the header, described above, cuts the crop and feeds it into the threshing cylinder. This consists of a series of horizontal *rasp bars* fixed across the path of the crop and in the shape of a quarter cylinder. Moving rasp bars or rub bars pull the crop through concaved grates that separate the grain and chaff from the straw. The grain heads fall through the fixed concaves. What happens next is dependent on the type of combine in question. In most modern combines, the grain is transported to the shoe by a set of 2, 3, or 4 (possibly more on the largest machines) augers, set parallel or semi-parallel to the rotor on axial mounted rotors and perpendicular Flow" combines.) In older Gleaner machines, these augers were not present. These combines are unique in that the cylinder and concave is set inside feederhouse instead of in the machine directly behind the feeder-house. Consequently, the material was moved by a "raddle chain" from underneath the concave to the walkers. The clean grain fell between the raddle and the walkers onto the shoe, while the straw, being longer and lighter, floated across onto the walkers to be expelled. On most other older machines, the cylinder was placed higher and farther back in the machine, and the grain moved to the shoe by falling down a "clean grain pan", and the straw "floated" across the concaves to the back of the walkers.

Since the Sperry-New Holland TR70 Twin-Rotor Combine came out in 1975, most manufacturers have combines with rotors in place of conventional cylinders. However, makers have now returned to the market with conventional models alongside their rotary line-up. A rotor is a long, longitudinally mounted rotating cylinder with plates similar to rub bars (except for in the above-mentioned Gleaner rotaries).

There are usually two sieves, one above the other. The sieves and basically a metal frame, that has many rows of "fingers" set reasonably close together. The angle of the fingers is adjustable as to change the clearance and control the size of material passing through. The top is set with more clearance than the bottom as to allow a gradual cleaning action. Setting the concave clearance, fan speed, and sieve size is critical to ensure that the crop is threshed properly, the grain is clean of debris, and that all of the grain entering the machine reaches the grain tank or 'hopper'. (Observe, for example, that when travelling uphill the fan speed must be reduced to account for the shallower gradient of the sieves.)

Heavy material, e.g., unthreshed heads, fall off the front of the sieves and are returned to the concave for re-threshing.

The straw walkers are located above the sieves, and also have holes in them. Any grain remaining attached to the straw is shaken off and falls onto the top sieve.

When the straw reaches the end of the walkers it falls out the rear of the combine. It can then be baled for cattle bedding or spread by two rotating straw spreaders with rubber arms. Most modern combines are equipped with a straw spreader.

Rotary and Conventional Designs

For some time, combine harvesters used the conventional design, which used a rotating cylinder at the front-end which knocked the seeds out of the heads, and then used the rest of the machine to separate the straw from the chaff, and the chaff from the grain. The TR70 from Sperry-New Holland was brought out in 1975 as the first rotary combine. Other manufacturers soon followed, International Harvester with their 'Axial Flow' in 1977 and Gleaner with their N6 in 1979.

IH McCormick 141 Combine ca. 1954-57

Tr85

In the decades before the widespread adoption of the rotary combine in the late seventies, several inventors had pioneered designs which relied more on centrifugal force for grain separation and less on gravity alone. By the early eighties, most major manufacturers had settled on a "walkerless" design with much larger threshing cylinders to do most of the work. Advantages were faster grain harvesting and gentler treatment of fragile seeds, which were often cracked by the faster rotational speeds of conventional combine threshing cylinders.

It was the disadvantages of the rotary combine (increased power requirements and over-pulverization of the straw by-product) which prompted a resurgence of conventional combines in the late nineties. Perhaps overlooked but nonetheless true, when the large engines that powered the rotary machines were employed in conventional machines, the two types of machines delivered similar production capacities. Also, research was beginning to show that incorporating above-ground crop residue (straw) into the soil is less useful for rebuilding soil fertility than previously believed. This meant that working pulverized straw into the soil became more of a hindrance than a benefit. An increase in feedlot beef production also created a higher demand for straw as fodder. Conventional combines, which use straw walkers, preserve the quality of straw and allow it to be baled and removed from the field.

Combine Fires

Grain combine fires are responsible for millions of dollars of loss each year. Fires usually start near the engine where dust and dry crop debris accumulate. Fires can also start when heat is introduced by bearings or gearboxes that have failed. From 1984 to 2000, 695 major grain combine fires were reported to U.S. local fire departments. Dragging chains to reduce static electricity was one method employed for preventing harvester fires, but the role of static electricity linked to causing harvester fires is yet to be established. The application of the appropriate synthetic greases will reduce the friction experienced at crucial points, i.e. chains, sprockets and gear boxes compared to petroleum based lubricants. Engines with synthetic lubricants will also remain significantly cooler during operation.

Reaper

Typical 20th century reaper, a tractor-drawn Fahr machine

A reaper is a farming tool or person that reaps (cuts and gathers) crops at harvest, when they are ripe.

Hand Reaping

A reaper cutting rye in Germany in 1949

Hand reaping is done by various means, including plucking the ears of grains directly by hand, cutting the grain stalks with a sickle, cutting them with a scythe, or a scythe fitted with a grain cradle. Reaping is usually distinguished from *mowing*, which uses similar implements, but is the traditional term for cutting grass for hay, rather than reaping crops.

The reaped grain stalks are gathered into *sheaves* (bunches), tied with string or with a twist of straw. Several sheaves (singular *sheaf*) are then leant against each other with the ears off the ground to dry out, forming a *stook*. After drying, the sheaves are gathered from the field and stacked, being placed with the ears inwards, then covered with thatch or a tarpaulin; this is called a *stack* or *rick*. In the British Isles a rick of sheaves is traditionally called a *corn rick*, to distinguish it from a *hay rick* ("corn" in British English means "grain", not "maize", which is not grown for grain there). Ricks are made in an area inaccessible to livestock, called a *rick-yard* or *stack-yard*. The corn-rick is later broken down and the sheaves threshed to separate the grain from the straw.

Collecting spilt grain from the field after reaping is called *gleaning*, and is traditionally done either by hand, or by penning animals such as chickens or pigs onto the field.

Hand reaping is now rarely done in industrialized countries, but is still the normal method where machines are unavailable or where access for them is limited (such as on narrow terraces).

The more or less skeletal figure of a reaper with a scythe – known as the "Grim Reaper" – is a common personification of death in many Western traditions and cultures. In this metaphor, death harvests the living, like a farmer harvests the crops.

Mechanical Reaping

A mechanical reaper or reaping machine is a mechanical, semi-automated device that harvests crops. Mechanical reapers are an important part of mechanized agriculture and a main feature of agricultural productivity.

Early History

Drawing of a "Gallic header"

It is believed that either Romans or the Belgae Gallics before them, invented a simple mechanical reaper that cut the ears without the straw and was pushed by oxen. This device was forgotten in the Dark Ages, during which period reapers reverted to using scythes and sickles to gather crops.

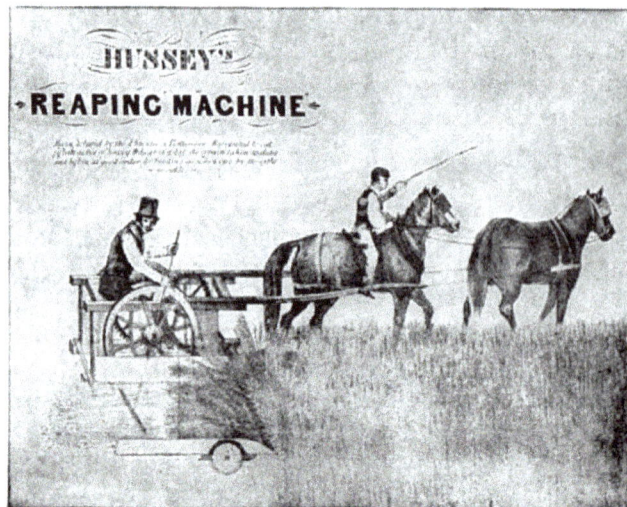

Hussey's reaping machine, 19th century

Thomas Dobbs (actor) of Birmingham invents a reaping machine in 1814, which consists of a circular saw or sickle, the grain is drawn or fed up to the saw by means of a pair of rollers.

Patrick Bell of Scotland created a reaper that used a revolving reel, cutting knife and canvas conveyor in 1828. This machine was used around his county and some may have been exported, but the device was never patented. One of Bell's reaping machines is preserved in the National Museum of Rural Life in Scotland.

Mechanical Reapers in the U.S.

The 19th century saw several inventors in the United States claim innovation in mechanical reapers.The various designs competed with each other, and were the subject of several lawsuits.

Obed Hussey in Ohio patented a reaper in 1833, the *Hussey Reaper*. Made in Baltimore, Maryland, Hussey's design was a major improvement in reaping efficiency. The new reaper only required two horses working in a non-strenuous manner, a man to work the machine, and another person to drive. In addition, the Hussey Reaper left an even and clean surface after its use.

McCormick's reaper at a presentation in Virginia

The *McCormick Reaper* was designed by Robert McCormick in Walnut Grove, Virginia. However, Robert became frustrated when he was unable to perfect his new device. His son Cyrus asked for permission to try to complete his father's project. With permission granted, the McCormick Reaper was patented by his son Cyrus McCormick in 1837 as a horse-drawn farm implement to cut small grain crops. This McCormick reaper machine had several special elements:

- a main wheel frame

- projected to the side a platform containing a cutter bar having fingers through which reciprocated a knife driven by a crank

- upon the outer end of the platform was a divider projecting ahead of the platform to separate the grain to be cut from that to be left standing

- a reel was positioned above the platform to hold the grain against the reciprocating knife to throw it back upon the platform

- the machine was drawn by a team walking at the side of the grain.

Cyrus McCormick claimed that his reaper was actually invented in 1831, giving him the true claim to the general design of the machine. Over the next few decades the Hussey and McCormick reapers would compete with each other in the marketplace, despite being quite similar.

In 1861, the United States Patent and Trademark Office issued a ruling on the invention of the polarizing reaper design. It was determined that the money made from reapers was in large part due to Obed Hussey. S.T. Shugert, the acting commissioner of patents, declared that Hussey's improvements were the foundation of their success. It was ruled that the heirs of Obed Hussey would be monetarily compensated for his hard work and innovation by those who had made money from the reaper. It was also ruled that McCormick's reaper patent would be renewed for another 7 years.

Although the McCormick reaper was a revolutionary innovation for the harvesting of crops, it did not experience mainstream success and acceptance until at least 20 years after it was patented by Cyrus McCormick. This was because the McCormick reaper lacked a quality unique to Obed Hussey's reaper. Hussey's reaper used a sawlike cutter bar that cut stalks far more effectively than McCormick's. Only once Cyrus McCormick was able to acquire the rights to Hussey's cutter-bar mechanism (around 1850) did a truly revolutionary machine emerge. Other factors in the gradual uptake of mechanized reaping included natural cultural conservatism among farmers (proven tradition versus new and unknown machinery); the poor state of many new farm fields, which were often littered with rocks, stumps, and areas of uneven soil, making the lifespan and operability of a reaping machine questionable; and some amount of fearful Luddism among farmers that the machine would take away jobs, most especially among hired manual labourers.

Even though McCormick is credited as the "inventor" of the mechanical reaper, he based his work on that of many others, including Roman, Scottish and American men, more than two decades of work by his father, and the aid of Jo Anderson, a slave held by his family.

Reapers in the Late 19th and 20th Century

Champion reaper, trade card from 1875

Horse-drawn reaper in Canada in 1941

After the first reapers were developed and patented, other slightly different reapers were distributed by several manufacturers throughout the world. The *Champion (Combined) Reapers and Mowers*, produced by the Champion Interest group (*Champion Machine Company*, later *Warder, Bushnell & Glessner*, absorbed in IHC 1902) in Springfield, Ohio in the second half of the 19th century, were highly successful in the 1880s in the United States. Springfield is still known as "The Champion City".

Generally, reapers developed into the 1872 invented reaper-binder, which reaped the crop and bound it into sheaves. By 1896, 400,000 reaper-binders were estimated to be harvesting grain and bananas. This was in turn replaced by the swather and eventually the combine harvester, which reaps and threshes in one operation.

In Central European agriculture reapers were – together with reaper-binders – common machines until the mid-20th century.

Threshing

Threshing is the process of loosening the edible part of cereal grain (or other crop) from the scaly, inedible chaff that surrounds it. It is the step in grain preparation after harvesting and before winnowing, which separates the loosened chaff from the grain. Threshing does not remove the bran from the grain.

An animal-powered thresher

Threshing may be done by beating the grain using a flail on a threshing floor. Another traditional method of threshing is to make donkeys or oxen walk in circles on the grain on a hard surface. A modern version of this in some areas is to spread the grain on the surface of a country road so the grain may be threshed by the wheels of passing vehicles.

Hand threshing was laborious, with a bushel of wheat taking about an hour. In the late 18th century, before threshing was mechanized, about one-quarter of agricultural labor was devoted to it.

Industrialization of threshing began in 1786 with the invention of the threshing machine by Scotsman Andrew Meikle. Today, in developed areas, it is now mostly done by machine, usually by a combine harvester, which harvests, threshes, and winnows the grain while it is still in the field.

The cereal may be stored in a threshing barn or silos.

A Threshing Bee is a festival held in communities to commemorate this process. The event is often held over multiple days and includes flea markets, hog wrestling, and dances.

Gallery

Threshing with hand flails, Great Britain, c. 1750. Image from c. 1875.

Irreler Bauertradition shows threshing by hand - Roscheider Hof Open Air Museum

Threshing floor, Santorini, Greece

Ludovic Bassarab's *La treierat* ("Threshing"), showing peasants in Romanian dress around a combine harvester

Threshing rice by hand (view in Full HD)

A farmer works on his field threshing with yolks in old age

Threshing with yaks in Astore, Gilgit-Baltistan

Wheat Threshing Demo at Goessel Threshing Days in Goessel, Kansas, 2010.

Video of a petrol-powered machine threshing rice in Hainan, China

Threshing of paddy by machine, Bangladesh

Threshing Machine

A threshing machine in operation

The thrashing machine, or, in modern spelling, threshing machine (or simply thresher), was first invented by Scottish mechanical engineer Andrew Meikle for use in agriculture. It was devised (c. 1786) for the separation of grain from stalks and husks. For thousands of years, grain was separated by hand with flails, and was very laborious and time-consuming, taking about one-quarter of agricultural labor by the 18th century. Mechanization of this process took much of the drudgery out of farm labour.

Early Social Impacts

Fig. 849. — *Batteuse Damey à manège direct placé sous la batteuse.*

Threshing machine from 1881

The Swing Riots in the UK were partly a result of the threshing machine. Following years of war, high taxes and low wages, farm labourers finally revolted in 1830. These farm labourers had faced unemployment for a number of years due to the widespread introduction of the threshing machine and the policy of enclosing fields. No longer were thousands of men needed to tend the crops, a few would suffice. With fewer jobs, lower wages and no prospects of things improving for these workers the threshing machine was the final straw, the machine was to place them on the brink of starvation. The Swing Rioters smashed threshing machines and threatened farmers who had them.

The riots were dealt with very harshly. Nine of the rioters were hanged and a further 450 were transported to Australia.

Later Adoption

Irreler Bauerntradition shows an early threshing machine (Stiftendrescher) at the Roscheider Hof Open Air Museum

Irreler Bauerntradition shows a winnowing machine (a forerunner of the threshing machine)
at the Roscheider Hof Open Air Museum

Early threshing machines were hand-fed and horse-powered. They were small by today's standards and were about the size of an upright piano. Later machines were steam-powered, driven by a portable engine or traction engine. Isaiah Jennings, a skilled inventor, created a small thresher that doesn't harm the straw in the process. In 1834, John Avery and Hiram Abial Pitts devised significant improvements to a machine that automatically threshes and separates grain from chaff, freeing farmers from a slow and laborious process. Avery and Pitts were granted United States patent #542 on December 29, 1837.

John Ridley, an Australian inventor, also developed a threshing machine in South Australia in 1843.

The 1881 *Household Cyclopedia* said of Meikle's machine:

> "Since the invention of this machine, Mr. Meikle and others have progressively introduced a variety of improvements, all tending to simplify the labour, and to augment the quantity of the work performed. When first erected, though the grain was equally well separated from the straw, yet as the whole of the straw, chaff, and grain, was indiscriminately thrown

into a confused heap, the work could only with propriety be considered as half executed. By the addition of rakes, or shakers, and two pairs of fanners, all driven by the same machinery, the different processes of thrashing, shaking, and winnowing are now all at once performed, and the grain immediately prepared for the public market. When it is added, that the quantity of grain gained from the superior powers of the machine is fully equal to a twentieth part of the crop, and that, in some cases, the expense of thrashing and cleaning the grain is considerably less than what was formerly paid for cleaning it alone, the immense saving arising from the invention will at once be seen.

"The expense of horse labour, from the increased value of the animal and the charge of his keeping, being an object of great importance, it is recommended that, upon all sizable farms, that is to say, where two hundred acres [800,000 m²], or upwards, of grain are sown, the machine should be worked by wind, unless where local circumstances afford the conveniency of water. Where coals are plenty and cheap, steam may be advantageously used for working the machine."

Steam-powered machines used belts connected to a traction engine; often both engine and thresher belonged to a contractor who toured the farms of a district. Steam remained a viable commercial option until the early post-WWII years.

Open-air museum in Saint-Hubert, Belgium.

Farming Process

Threshing is just one process in getting cereals to the grinding mill and customer. The wheat needs to be grown, cut, stooked (shocked, bundled), hauled, threshed, de-chaffed, straw baled, and then the grain hauled to a grain elevator. For many years each of these steps was an individual process, requiring teams of workers and many machines. In the steep hill wheat country of Palouse in the Northwest of the United States, steep ground meant moving machinery around was problematic and prone to rolling. To reduce the amount of work on the sidehills, the idea arose of combining the wheat binder and thresher into one machine, known as a combine harvester. About 1910, horse pulled combines appeared and became a success. Later, gas and diesel engines appeared with other refinements and specifications.

Modern Developments

In Europe and Americas

Threshing of paddy by machine, Bangladesh.

Modern day combine harvesters (or simply combines) operate on the same principles and use the same components as the original threshing machines built in the 19th century. Combines also perform the reaping operation at the same time. The name *combine* is derived from the fact that the two steps are combined in a single machine. Also, most modern combines are self-powered (usually by a diesel engine) and self-propelled, although tractor powered, pull type combines models were offered by John Deere and Case International into the 1990s.

Today, as in the 19th century, the threshing begins with a cylinder and concave. The cylinder has sharp serrated bars, and rotates at high speed (about 500 RPM), so that the bars beat against the grain. The concave is curved to match the curve of the cylinder, and serves to hold the grain as it is beaten. The beating releases the grain from the straw and chaff.

Whilst the majority of the grain falls through the concave, the straw is carried by a set of "walkers" to the rear of the machine, allowing any grain and chaff still in the straw to fall below. Below the straw walkers, a fan blows a stream of air across the grain, removing dust and fines and blowing them away.

The grain, either coming through the concave or the walkers, meets a set of sieves mounted on an assembly called a shoe, which is shaken mechanically. The top sieve has larger openings, and serves to remove large pieces of chaff from the grain. The lower sieve separates clean grain, which falls through, from incompletely threshed pieces. The incompletely threshed grain is returned to the cylinder by means of a system of conveyors, where the process repeats.

Some threshing machines were equipped with a bagger, which invariably held two bags, one being filled, and the other being replaced with an empty. A worker called a *sewer* removed and replaced the bags, and sewed full bags shut with a needle and thread. Other threshing machines would discharge grain from a conveyor, for bagging by hand. Combines are equipped with a grain tank, which accumulates grain for deposit in a truck or wagon.

A large amount of chaff and straw would accumulate around a threshing machine, and several innovations, such as the air chaffer, were developed to deal with this. Combines generally chop and disperse straw as they move through the field, though the chopping is disabled when the straw is to be baled, and chaff collectors are sometimes used to prevent the dispersal of weed seed throughout a field.

The corn sheller was almost identical in design, with slight modifications to deal with the larger kernel size and presence of cobs. Modern-day combines can be adjusted to work with any grain crop, and many unusual seed crops.

Both the older and modern machines require a good deal of skill to operate. The concave clearance, cylinder speed, fan velocity, sieve sizes, and feeding rate must be adjusted for crop conditions.

Another development in Asia

Video of a petrol-powered machine threshing rice in Hainan, China

From the early 20th century, petrol or diesel-powered threshing machines, designed especially to thresh rice, the most important crop in Asia, have been developed along different lines to the modern combine.

Even after the combine was invented and became popular, a new compact-size thresher called a *harvester*, with wheels, still remains in use and at present it is available from a Japanese agricultural manufacturer. The compact-size machine is very convenient to handle in small terrace fields in mountain areas where a large machine, such as combine, is not usable.

People there use this harvester with a modern compact binder.

Preservation

A number of older threshing machines have survived into preservation. They are often to be seen in operation at live steam festivals and traction engine rallies such as the Great Dorset Steam Fair in England, and the Western Minnesota Steam Threshers Reunion in northwest Minnesota.

Musical references

Irish songwriter John Duggan immortalised the threshing machine in a song *The Old Thrashing Mill*. The song has been recorded by Foster and Allen and Brendan Shine.

On the Alan Lomax collection Songs of Seduction (Rounder Select, 2000), there's a bawdy Irish folk song called "The Thrashing Machine" sung by tinker Annie O'Neil, as recorded in the early 20th Century.

In his film score for "Of Mice and Men" (1939) and consequently in his collection "Music for the Movies" (1942), American composer Aaron Copland titled a section of the score "Threshing Machines," to suit a scene in the Lewis Milestone film where Curley is threatening Slim over giving May a puppy, when many of the itinerant worker men are standing around or working on threshers.

In the song Thrasher from the album Rust Never Sleeps, Neil Young compares the modern threshing machine's technique of separating wheat from wheat stalks to the natural forces of time that separate close friends from one another.

Threshing machines appear in Twenty One Pilots' music video for the song House of Gold.

Winnowing

Rice winnowing, Uttarakhand, India

Winnowing in a Dalit village in Tamil Nadu, India

Use of winnowing forks

Wind winnowing is an agricultural method developed by ancient cultures for separating grain from chaff. It is also used to remove weevils or other pests from stored grain. Threshing, the loosening of grain or seeds from the husks and straw, is the step in the chaff-removal process that comes before winnowing.

In its simplest form it involves throwing the mixture into the air so that the wind blows away the lighter chaff, while the heavier grains fall back down for recovery. Techniques included using a winnowing fan (a shaped basket shaken to raise the chaff) or using a tool (a winnowing fork or shovel) on a pile of harvested grain.

In Greek Culture

The winnowing-fan (λικνον [*líknon*], also meaning a "cradle") featured in the rites accorded Dionysus and in the Eleusinian Mysteries: "it was a simple agricultural implement taken over and mysticised by the religion of Dionysus," Jane Ellen Harrison remarked. *Dionysus Liknites* ("Dionysus of the winnowing fan") was wakened by the Dionysian women, in this instance called *Thyiades*, in a cave on Parnassus high above Delphi; the winnowing-fan links the god connected with the mystery religions to the agricultural cycle, but mortal Greek babies too were laid in a winnowing-fan. In Callimachus' *Hymn to Zeus*, Adrasteia lays the infant Zeus in a golden *líknon*, her goat suckles him and he is given honey.

In the Odyssey, the dead oracle Teiresias tells Odysseus to walk away from Ithaca with an oar until a wayfarer tells him it is a winnowing fan (i.e., until Odysseus has come so far from the sea that people don't recognize oars), and there to build a shrine to Poseidon.

In China

Chinese rotary fan winnowing machine, from the *Tiangong Kaiwu* encyclopedia (1637)

In Ancient China the method was improved by mechanisation with the development of the rotary winnowing fan, which used a cranked fan to produce the airstream. This was featured in Wang Zhen's book the *Nong Shu* of 1313 AD.

In the Old Testament

In the Old Testament the word winnow is used in several verses in different books in the New International Version while other versions of the bible translate the action as "fan", "throw" or the separating tool as "pitchfork", "shovel", "winnowing fan", or "winnowing instrument".

Ruth 3:2 "Now Boaz, with whose women you have worked, is a relative of ours. Tonight he will be winnowing barley on the threshing floor."

Proverbs 20:8 "When a king sits on his throne to judge, he winnows out all evil with his eyes."

Proverbs 20:26 "A wise king winnows out the wicked; he drives the threshing wheel over them."

Isaiah 41:16 "You will winnow them, the wind will pick them up, and a gale will blow them away. But you will rejoice in the Lord and glory in the Holy One of Israel."

Jeremiah 4:11 "At that time this people and Jerusalem will be told, "A scorching wind from the barren heights in the desert blows toward my people, but not to winnow or cleanse;"

Jeremiah 15:7 "I will winnow them with a winnowing fork at the city gates of the land. I will bring bereavement and destruction on my people, for they have not changed their ways."

Jeremiah 51:2 "I will send foreigners to Babylon to winnow her and to devastate her land; they will oppose her on every side in the day of her disaster."

In the New Testament

In Matthew 3:12, a sentence introduces the separation of wheat and chaff (good and bad) by "His winnowing fan is in his hand" (American Standard Bible and New American Bible translation). The New International Version, the New Revised Standard Version and the New American Standard Version translate the term as "winnowing fork."

In Europe

Le vanneur (*The Winnower*) by Jean-François Millet, a 19th-century depiction of winnowing by fan

In Saxon settlements such as one identified in Northumberland as Bede's Ad Gefrin (now called Yeavering) the buildings were shown by an excavator's reconstruction to have opposed entries. In barns a draught created by the use of these opposed doorways was used in winnowing.

The technique developed by the Chinese was not adopted in Europe until the 18th century, when winnowing machines used a 'sail fan'. The rotary winnowing fan was exported to Europe, brought there by Dutch sailors between 1700 and 1720. Apparently they had obtained them from the Dutch settlement of Batavia in Java, Dutch East Indies. The Swedes imported some from south China at about the same time and Jesuits had taken several to France from China by 1720. Until the beginning of the 18th century, no rotary winnowing fans existed in the West.

In the United States

The development of the winnowing barn allowed rice plantations in South Carolina to increase their yields dramatically.

Mechanization of the Process

Winnowing machine from 1839

In 1737 Andrew Rodger, a farmer on the estate of Cavers in Roxburghshire, developed a winnowing machine for corn, called a 'Fanner'. These were successful and the family sold them throughout Scotland for many years. Some Scottish Presbyterian ministers saw the fanners as sins against God, for wind was a thing specially made by him and an artificial wind was a daring and impious attempt to usurp what belonged to God alone. As the Industrial Revolution, the winnowing process was mechanized by the invention of additional winnowing machines, such as fanning mills.

Postharvest

Drying and bagging of peanuts in Jiangxia District, Hubei, China

In agriculture, postharvest handling is the stage of crop production immediately following harvest, including cooling, cleaning, sorting and packing. The instant a crop is removed from the ground, or separated from its parent plant, it begins to deteriorate. Postharvest treatment largely determines final quality, whether a crop is sold for fresh consumption, or used as an ingredient in a processed food product.

Goals

Drying chili peppers. Milyanfan, Kyrgyzstan

The most important goals of post-harvest handling are keeping the product cool, to avoid moisture loss and slow down undesirable chemical changes, and avoiding physical damage such as bruising, to delay spoilage. Sanitation is also an important factor, to reduce the possibility of pathogens that could be carried by fresh produce, for example, as residue from contaminated washing water.

After the field, post-harvest processing is usually continued in a packing house. This can be a simple shed, providing shade and running water, or a large-scale, sophisticated, mechanised facility, with conveyor belts, automated sorting and packing stations, walk-in coolers and the like. In mechanised harvesting, processing may also begin as part of the actual harvest process, with initial cleaning and sorting performed by the harvesting machinery.

Initial post-harvest storage conditions are critical to maintaining quality. Each crop has an optimum range of storage temperature and humidity. Also, certain crops cannot be effectively stored together, as unwanted chemical interactions can result. Various methods of high-speed cooling, and sophisticated refrigerated and atmosphere-controlled environments, are employed to prolong freshness, particularly in large-scale operations.

Regardless of the scale of harvest, from domestic garden to industrialised farm, the basic principles of post-harvest handling for most crops are the same: handle with care to avoid damage (cutting, crushing, bruising), cool immediately and maintain in cool conditions, and cull (remove damaged items).

Postharvest Shelf Life

Once harvested, vegetable and fruit are subject to the active process of senescence. Numerous biochemical processes continuously change the original composition of the crop until it becomes unmarketable. The period during which consumption is considered acceptable is defined as the time of "postharvest shelf life".

Postharvest shelf life is typically determined by objective methods that determine the overall appearance, taste, flavour, and texture of the commodity. These methods usually include a combination of sensorial, biochemical, mechanical, and colorimetric (optical) measurements. A recent study attempted (and failed) to discover a biochemical marker and fingerprint methods as indices for freshness.

Postharvest Physiology

Postharvest physiology is the scientific study of the physiology of living plant tissues after they have denied further nutrition by picking. It has direct applications to postharvest handling in establishing the storage and transport conditions that best prolong shelf life.

An example of the importance of the field to post-harvest handling is the discovery that ripening of fruit can be delayed, and thus their storage prolonged, by preventing fruit tissue respiration. This insight allowed scientists to bring to bear their knowledge of the fundamental principles and mechanisms of respiration, leading to post-harvest storage techniques such as cold storage, gaseous storage, and waxy skin coatings. Another well-known example is the finding that ripening may be brought on by treatment with ethylene.

References

- Constable, George; Somerville, Bob (2003). A Century of Innovation: Twenty Engineering Achievements That Transformed Our Lives, Chapter 7, Agricultural Mechanization. Washington, DC: Joseph Henry Press. ISBN 0-309-08908-5.

- Daniel, Gross; Forbes Magazine Staff (August 1997). Greatest Business Stories of All Time (First ed.). New York: John Wiley & Sons, Inc. p. 27. ISBN 0-471-19653-3.

- Pripps, Robert N.; Morland, Andrew (photographer) (1993), Farmall Tractors: History of International McCormick-Deering Farmall Tractors, Farm Tractor Color History Series, Osceola, WI, USA: MBI, ISBN 978-0-87938-763-1, p. 17.

- Atack, Jeremy; Passell, Peter (1994). A New Economic View of American History. New York: W.W. Norton and Co. pp. 282–3. ISBN 0-393-96315-2.

- Clark, Gregory (2007). A Farewell to Alms: A Brief Economic History of the World. Princeton University Press. p. 286. ISBN 978-0-691-12135-2.

Irrigation

Agriculture on the whole depends on natural and man-made systems of irrigation. The creation of canals and aqueducts has been going on for centuries. Irrigation is very important for farming. Some methods discussed in this chapter include surface irrigation and drip irrigation. The major categories of irrigation are dealt with in great details in the chapter.

Irrigation

An irrigation sprinkler watering a lawn

Irrigation is the method in which water is supplied to plants at regular intervals for agriculture. It is used to assist in the growing of agricultural crops, maintenance of landscapes, and revegetation of disturbed soils in dry areas and during periods of inadequate rainfall. Additionally, irrigation also has a few other uses in crop production, which include protecting plants against frost, suppressing weed growth in grain fields and preventing soil consolidation. In contrast, agriculture that relies only on direct rainfall is referred to as rain-fed or dry land farming.

Irrigation systems are also used for dust suppression, disposal of sewage, and in mining. Irrigation is often studied together with drainage, which is the natural or artificial removal of surface and sub-surface water from a given area.

Irrigation has been a central feature of agriculture for over 5,000 years and is the product of many cultures. Historically, it was the basis for economies and societies across the globe, from Asia to the Southwestern United States.

Irrigation canal in Osmaniye, Turkey

History

Animal-powered irrigation, Upper Egypt, ca. 1846

An example of an irrigation system common on the Indian subcontinent.
Artistic impression on the banks of Dal Lake, Kashmir, India

Inside a karez tunnel at Turpan, Sinkiang

Archaeological investigation has identified as evidence of irrigation where the natural rainfall was insufficient to support crops for rainfed agriculture.

Perennial irrigation was practiced in the Mesopotamian plain whereby crops were regularly watered throughout the growing season by coaxing water through a matrix of small channels formed in the field.

irrigation in Tamil Nadu (India)

Ancient Egyptians practiced *Basin irrigation* using the flooding of the Nile to inundate land plots which had been surrounded by dykes. The flood water was held until the fertile sediment had settled before the surplus was returned to the watercourse. There is evidence of the ancient Egyptian pharaoh Amenemhet III in the twelfth dynasty (about 1800 BCE) using the natural lake of the Faiyum Oasis as a reservoir to store surpluses of water for use during the dry seasons, the lake swelled annually from flooding of the Nile.

The Ancient Nubians developed a form of irrigation by using a waterwheel-like device called a *sakia*. Irrigation began in Nubia some time between the third and second millennium BCE. It largely depended upon the flood waters that would flow through the Nile River and other rivers in what is now the Sudan.

In sub-Saharan Africa irrigation reached the Niger River region cultures and civilizations by the first or second millennium BCE and was based on wet season flooding and water harvesting.

Terrace irrigation is evidenced in pre-Columbian America, early Syria, India, and China. In the Zana Valley of the Andes Mountains in Peru, archaeologists found remains of three irrigation canals radiocarbon dated from the 4th millennium BCE, the 3rd millennium BCE and the 9th century CE. These canals are the earliest record of irrigation in the New World. Traces of a canal possibly dating from the 5th millennium BCE were found under the 4th millennium canal. Sophisticated irrigation and storage systems were developed by the Indus Valley Civilization in present-day Pakistan and North India, including the reservoirs at Girnar in 3000 BCE and an early canal irrigation system from circa 2600 BCE. Large scale agriculture was practiced and an extensive network of canals was used for the purpose of irrigation.

Ancient Persia (modern day Iran) as far back as the 6th millennium BCE, where barley was grown in areas where the natural rainfall was insufficient to support such a crop. The Qanats, developed in ancient Persia in about 800 BCE, are among the oldest known irrigation methods still in use today. They are now found in Asia, the Middle East and North Africa. The system comprises a network of vertical wells and gently sloping tunnels driven into the sides of cliffs and steep hills to tap groundwater. The noria, a water wheel with clay pots around the rim powered by the flow of the stream (or by animals where the water source was still), was first brought into use at about this time, by Roman settlers in North Africa. By 150 BCE the pots were fitted with valves to allow smoother filling as they were forced into the water.

The irrigation works of ancient Sri Lanka, the earliest dating from about 300 BCE, in the reign of King Pandukabhaya and under continuous development for the next thousand years, were one of the most complex irrigation systems of the ancient world. In addition to underground canals, the Sinhalese were the first to build completely artificial reservoirs to store water. Due to their engineering superiority in this sector, they were often called 'masters of irrigation'. Most of these irrigation systems still exist undamaged up to now, in Anuradhapura and Polonnaruwa, because of the advanced and precise engineering. The system was extensively restored and further extended during the reign of King Parakrama Bahu (1153–1186 CE).

China

The oldest known hydraulic engineers of China were Sunshu Ao (6th century BCE) of the Spring and Autumn Period and Ximen Bao (5th century BCE) of the Warring States period, both of whom worked on large irrigation projects. In the Sichuan region belonging to the State of Qin of ancient China, the Dujiangyan Irrigation System was built in 256 BCE to irrigate an enormous area of farmland that today still supplies water. By the 2nd century AD, during the Han Dynasty, the Chinese also used chain pumps that lifted water from lower elevation to higher elevation. These were powered by manual foot pedal, hydraulic waterwheels, or rotating mechanical wheels pulled by oxen. The water was used for public works of providing water for urban residential quarters and palace gardens, but mostly for irrigation of farmland canals and channels in the fields.

Korea

In 15th century Korea, the world's first rain gauge, *uryanggye* (Korean:우량계), was invented in 1441. The inventor was Jang Yeong-sil, a Korean engineer of the Joseon Dynasty, under the active direction of the king, Sejong the Great. It was installed in irrigation tanks as part of a nationwide system to measure and collect rainfall for agricultural applications. With this instrument, planners and farmers could make better use of the information gathered in the survey.

North America

In North America, the Hohokam were the only culture to rely on irrigation canals to water their crops, and their irrigation systems supported the largest population in the Southwest by AD 1300. The Hohokam constructed an assortment of simple canals combined with weirs in their various agricultural pursuits. Between the 7th and 14th centuries, they also built and maintained extensive irrigation networks along the lower Salt and middle Gila rivers that rivaled the complexity of those used in the ancient Near East, Egypt, and China. These were constructed using relatively simple excavation tools, without the benefit of advanced engineering technologies, and achieved drops of a few feet per mile, balancing erosion and siltation. The Hohokam cultivated varieties of cotton, tobacco, maize, beans and squash, as well as harvested an assortment of wild plants. Late in the Hohokam Chronological Sequence, they also used extensive dry-farming systems, primarily to grow agave for food and fiber. Their reliance on agricultural strategies based on canal irrigation, vital in their less than hospitable desert environment and arid climate, provided the basis for the aggregation of rural populations into stable urban centers.

Present Extent

Irrigation ditch in Montour County, Pennsylvania, off Strawberry Ridge Road

In the mid-20th century, the advent of diesel and electric motors led to systems that could pump groundwater out of major aquifers faster than drainage basins could refill them. This can lead to permanent loss of aquifer capacity, decreased water quality, ground subsidence, and other problems. The future of food production in such areas as the North China Plain, the Punjab, and the Great Plains of the US is threatened by this phenomenon.

At the global scale, 2,788,000 km² (689 million acres) of fertile land was equipped with irrigation infrastructure around the year 2000. About 68% of the area equipped for irrigation is located in Asia, 17% in the Americas, 9% in Europe, 5% in Africa and 1% in Oceania. The largest contiguous areas of high irrigation density are found:

- In Northern India and Pakistan along the Ganges and Indus rivers

- In the Hai He, Huang He and Yangtze basins in China

- Along the Nile river in Egypt and Sudan

- In the Mississippi-Missouri river basin and in parts of California

Smaller irrigation areas are spread across almost all populated parts of the world.

Only eight years later in 2008, the scale of irrigated land increased to an estimated total of 3,245,566 km² (802 million acres), which is nearly the size of India.

Types of Irrigation

Irrigation of land in Punjab, Pakistan

Various types of irrigation techniques differ in how the water obtained from the source is distributed within the field. In general, the goal is to supply the entire field uniformly with water, so that each plant has the amount of water it needs, neither too much nor too little.

Surface Irrigation

In *surface* (*furrow*, *flood*, or *level basin*) irrigation systems, water moves across the surface of agricultural lands, in order to wet it and infiltrate into the soil. Surface irrigation can be subdivided into furrow, *borderstrip or basin irrigation*. It is often called *flood irrigation* when the irrigation results in flooding or near flooding of the cultivated land. Historically, this has been the most common method of irrigating agricultural land and still is in most parts of the world.

Where water levels from the irrigation source permit, the levels are controlled by dikes, usually plugged by soil. This is often seen in terraced rice fields (rice paddies), where the method is used to flood or control the level of water in each distinct field. In some cases, the water is pumped, or lifted by human or animal power to the level of the land. The field water efficiency of surface irrigation is typically lower than other forms of irrigation but has the potential for efficiencies in the range of 70% - 90% under appropriate management.

Localized Irrigation

Impact sprinkler head

Localized irrigation is a system where water is distributed under low pressure through a piped network, in a pre-determined pattern, and applied as a small discharge to each plant or adjacent to it. Drip irrigation, spray or micro-sprinkler irrigation and bubbler irrigation belong to this category of irrigation methods.

Subsurface Textile Irrigation

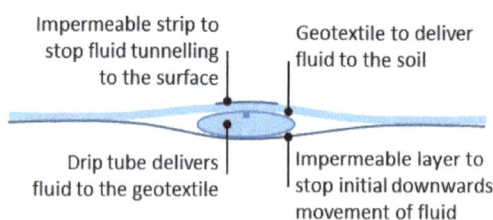

Impermeable strip to stop fluid tunnelling to the surface

Geotextile to deliver fluid to the soil

Drip tube delivers fluid to the geotextile

Impermeable layer to stop initial downwards movement of fluid

Diagram showing the structure of an example SSTI installation

Subsurface Textile Irrigation (SSTI) is a technology designed specifically for subsurface irrigation in all soil textures from desert sands to heavy clays. A typical subsurface textile irrigation system has an impermeable base layer (usually polyethylene or polypropylene), a drip line running along that base, a layer of geotextile on top of the drip line and, finally, a narrow impermeable layer on top of the geotextile (see diagram). Unlike standard drip irrigation, the spacing of emitters in the drip pipe is not critical as the geotextile moves the water along the fabric up to 2 m from the dripper.

Drip Irrigation

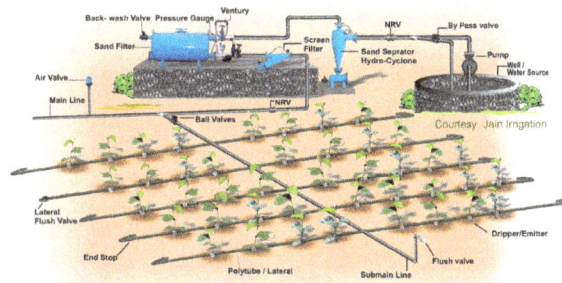

Drip irrigation layout and its parts

Grapes in Petrolina, only made possible in this semi arid area by drip irrigation

Drip (or micro) irrigation, also known as trickle irrigation, functions as its name suggests. In this system water falls drop by drop just at the position of roots. Water is delivered at or near the root zone of plants, drop by drop. This method can be the most water-efficient method of irrigation, if managed properly, since evaporation and runoff are minimized. The field water efficiency of drip irrigation is typically in the range of 80 to 90 percent when managed correctly.

In modern agriculture, drip irrigation is often combined with plastic mulch, further reducing evaporation, and is also the means of delivery of fertilizer. The process is known as *fertigation*.

Deep percolation, where water moves below the root zone, can occur if a drip system is operated for too long or if the delivery rate is too high. Drip irrigation methods range from very high-tech and computerized to low-tech and labor-intensive. Lower water pressures are usually needed than for most other types of systems, with the exception of low energy center pivot systems and surface irrigation systems, and the system can be designed for uniformity throughout a field or for precise water delivery to individual plants in a landscape containing a mix of plant species. Although it is difficult to regulate pressure on steep slopes, pressure compensating emitters are available, so the field does not have to be level. High-tech solutions involve precisely calibrated emitters located along lines of tubing that extend from a computerized set of valves.

Irrigation Using Sprinkler Systems

Sprinkler irrigation of blueberries in Plainville, New York, United States

A traveling sprinkler at Millets Farm Centre, Oxfordshire, United Kingdom

In *sprinkler* or overhead irrigation, water is piped to one or more central locations within the field and distributed by overhead high-pressure sprinklers or guns. A system utilizing sprinklers, sprays, or guns mounted overhead on permanently installed risers is often referred to as a *solid-set* irrigation system. Higher pressure sprinklers that rotate are called *rotors* an are driven by a ball drive, gear drive, or impact mechanism. Rotors can be designed to rotate in a full or partial circle. Guns are similar to rotors, except that they generally operate at very high pressures of 40 to 130 lbf/in^2 (275 to 900 kPa) and flows of 50 to 1200 US gal/min (3 to 76 L/s), usually with nozzle diameters in the range of 0.5 to 1.9 inches (10 to 50 mm). Guns are used not only for irrigation, but also for industrial applications such as dust suppression and logging.

Sprinklers can also be mounted on moving platforms connected to the water source by a hose. Automatically moving wheeled systems known as *traveling sprinklers* may irrigate areas such as small farms, sports fields, parks, pastures, and cemeteries unattended. Most of these utilize a length of polyethylene tubing wound on a steel drum. As the tubing is wound on the drum powered by the irrigation water or a small gas engine, the sprinkler is pulled across the field. When the sprinkler arrives back at the reel the system shuts off. This type of system is known to most people as a "waterreel" traveling irrigation sprinkler and they are used extensively for dust suppression, irrigation, and land application of waste water.

Other travelers use a flat rubber hose that is dragged along behind while the sprinkler platform is pulled by a cable. These cable-type travelers are definitely old technology and their use is limited in today's modern irrigation projects.

Irrigation Using Center Pivot

A small center pivot system from beginning to end

The hub of a center-pivot irrigation system

Rotator style pivot applicator sprinkler

Center pivot with drop sprinklers

Wheel line irrigation system in Idaho, 2001

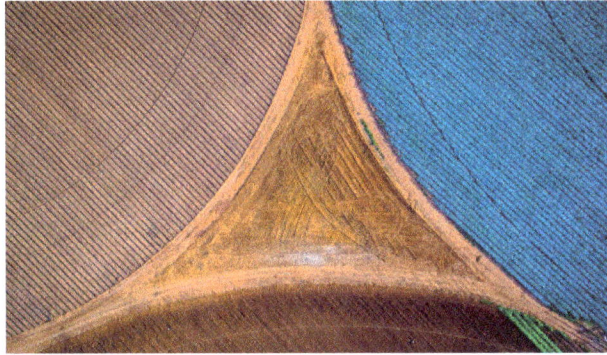

Center pivot irrigation

Center pivot irrigation is a form of sprinkler irrigation consisting of several segments of pipe (usually galvanized steel or aluminium) joined together and supported by trusses, mounted on wheeled towers with sprinklers positioned along its length. The system moves in a circular pattern and is fed with water from the pivot point at the center of the arc. These systems are found and used in all parts of the world and allow irrigation of all types of terrain. Newer systems have drop sprinkler heads as shown in the image that follows.

Most center pivot systems now have drops hanging from a u-shaped pipe attached at the top of the pipe with sprinkler head that are positioned a few feet (at most) above the crop, thus limiting evaporative losses. Drops can also be used with drag hoses or bubblers that deposit the water directly on the ground between crops. Crops are often planted in a circle to conform to the center pivot. This type of system is known as LEPA (Low Energy Precision Application). Originally, most center pivots were water powered. These were replaced by hydraulic systems (*T-L Irrigation*) and electric motor driven systems (Reinke, Valley, Zimmatic). Many modern pivots feature GPS devices.

Irrigation by Lateral Move (Side Roll, Wheel Line, Wheelmove)

A *series of pipes, each with a wheel* of about 1.5 m diameter permanently affixed to its midpoint, and sprinklers along its length, are coupled together. Water is supplied at one end using a large hose. After sufficient irrigation has been applied to one strip of the field, the hose is removed, the water drained from the system, and the assembly rolled either by hand or with a purpose-built mechanism, so that the sprinklers are moved to a different position across the field. The hose is reconnected. The process is repeated in a pattern until the whole field has been irrigated.

This system is less expensive to install than a center pivot, but much more labor-intensive to operate - it does not travel automatically across the field: it applies water in a stationary strip, must be drained, and then rolled to a new strip. Most systems use 4 or 5-inch (130 mm) diameter aluminum pipe. The pipe doubles both as water transport and as an axle for rotating all the wheels. A drive system (often found near the centre of the wheel line) rotates the clamped-together pipe sections as a single axle, rolling the whole wheel line. Manual adjustment of individual wheel positions may be necessary if the system becomes misaligned.

Wheel line systems are limited in the amount of water they can carry, and limited in the height of crops that can be irrigated. One useful feature of a lateral move system is that it consists of sections that can be easily disconnected, adapting to field shape as the line is moved. They are most often used for small, rectilinear, or oddly-shaped fields, hilly or mountainous regions, or in regions where labor is inexpensive.

Sub-irrigation

Subirrigation has been used for many years in field crops in areas with high water tables. It is a method of artificially raising the water table to allow the soil to be moistened from below the plants' root zone. Often those systems are located on permanent grasslands in lowlands or river valleys and combined with drainage infrastructure. A system of pumping stations, canals, weirs and gates allows it to increase or decrease the water level in a network of ditches and thereby control the water table.

Sub-irrigation is also used in commercial greenhouse production, usually for potted plants. Water is delivered from below, absorbed upwards, and the excess collected for recycling. Typically, a solution of water and nutrients floods a container or flows through a trough for a short period of time, 10–20 minutes, and is then pumped back into a holding tank for reuse. Sub-irrigation in greenhouses requires fairly sophisticated, expensive equipment and management. Advantages are water and nutrient conservation, and labor-saving through lowered system maintenance and automation. It is similar in principle and action to subsurface basin irrigation.

Irrigation Automatically, Non-electric Using Buckets and Ropes

Besides the common manual watering by bucket, an automated, natural version of this also exists. Using plain polyester ropes combined with a prepared ground mixture can be used to water plants from a vessel filled with water.

The ground mixture would need to be made depending on the plant itself, yet would mostly consist of black potting soil, vermiculite and perlite. This system would (with certain crops) allow to save expenses as it does not consume any electricity and only little water (unlike sprinklers, water timers, etc.). However, it may only be used with certain crops (probably mostly larger crops that do not need a humid environment; perhaps e.g. paprikas).

Irrigation Using Water Condensed From Humid Air

In countries where at night, humid air sweeps the countryside.Water can be obtained from the humid air by condensation onto cold surfaces. This is for example practiced in the vineyards at Lanzarote using stones to condense water or with various fog collectors based on canvas or foil sheets.

In-ground Irrigation

Most commercial and residential irrigation systems are "in ground" systems, which means that everything is buried in the ground. With the pipes, sprinklers, emitters (drippers), and irrigation valves being hidden, it makes for a cleaner, more presentable landscape without garden hoses or other items having to be moved around manually. This does, however, create some drawbacks in the maintenance of a completely buried system.

Most irrigation systems are divided into zones. A zone is a single irrigation valve and one or a group of drippers or sprinklers that are connected by pipes or tubes. Irrigation systems are divided into zones because there is usually not enough pressure and available flow to run sprinklers for an entire yard or sports field at once. Each zone has a solenoid valve on it that is controlled via wire by an irrigation controller. The irrigation controller is either a mechanical (now the "dinosaur" type) or electrical device that signals a zone to turn on at a specific time and keeps it on for a specified amount of time. "Smart Controller" is a recent term for a controller that is capable of adjusting the watering time by itself in response to current environmental conditions. The smart controller determines current conditions by means of historic weather data for the local area, a soil moisture sensor (water potential or water content), rain sensor, or in more sophisticated systems satellite feed weather station, or a combination of these.

When a zone comes on, the water flows through the lateral lines and ultimately ends up at the irrigation emitter (drip) or sprinkler heads. Many sprinklers have pipe thread inlets on the bottom of them which allows a fitting and the pipe to be attached to them. The sprinklers are usually installed with the top of the head flush with the ground surface. When the water is pressurized, the head will pop up out of the ground and water the desired area until the valve closes and shuts off that zone. Once there is no more water pressure in the lateral line, the sprinkler head will retract back into the ground. Emitters are generally laid on the soil surface or buried a few inches to reduce evaporation losses.

Water Sources

Irrigation is underway by pump-enabled extraction directly from the Gumti, seen in the background, in Comilla, Bangladesh.

Irrigation water can come from groundwater (extracted from springs or by using wells), from surface water (withdrawn from rivers, lakes or reservoirs) or from non-conventional sources like treated wastewater, desalinated water or drainage water. A special form of irrigation using surface water is spate irrigation, also called floodwater harvesting. In case of a flood (spate), water is diverted to normally dry river beds (wadis) using a network of dams, gates and channels and spread over large areas. The moisture stored in the soil will be used thereafter to grow crops. Spate irrigation areas are in particular located in semi-arid or arid, mountainous regions. While floodwater harvesting belongs to the accepted irrigation methods, rainwater harvesting is usually not considered as a form of irrigation. Rainwater harvesting is the collection of runoff water from roofs or unused land and the concentration of this.

Around 90% of wastewater produced globally remains untreated, causing widespread water pollution, especially in low-income countries. Increasingly, agriculture uses untreated wastewater as a source of irrigation water. Cities provide lucrative markets for fresh produce, so are attractive to farmers. However, because agriculture has to compete for increasingly scarce water resources with industry and municipal users, there is often no alternative for farmers but to use water polluted with urban waste, including sewage, directly to water their crops. Significant health hazards can result from using water loaded with pathogens in this way, especially if people eat raw vegetables that have been irrigated with the polluted water. The International Water Management Institute has worked in India, Pakistan, Vietnam, Ghana, Ethiopia, Mexico and other countries on various projects aimed at assessing and reducing risks of wastewater irrigation. They advocate a 'multiple-barrier' approach to wastewater use, where farmers are encouraged to adopt various risk-reducing behaviours. These include ceasing irrigation a few days before harvesting to allow pathogens to die off in the sunlight, applying water carefully so it does not contaminate leaves likely to be eaten raw, cleaning vegetables with disinfectant or allowing fecal sludge used in farming to dry before being used as a human manure. The World Health Organization has developed guidelines for safe water use.

There are numerous benefits of using recycled water for irrigation, including the low cost (when compared to other sources, particularly in an urban area), consistency of supply (regardless of season, climatic conditions and associated water restrictions), and general consistency of quality. Irrigation of recycled wastewater is also considered as a means for plant fertilization and particularly nutrient supplementation. This approach carries with it a risk of soil and water pollution through excessive wastewater application. Hence, a detailed understanding of soil water conditions is essential for effective utilization of wastewater for irrigation.

Efficiency

Young engineers restoring and developing the old Mughal irrigation system during the reign of the Mughal Emperor Bahadur Shah II

Modern irrigation methods are efficient enough to supply the entire field uniformly with water, so that each plant has the amount of water it needs, neither too much nor too little. Water use efficiency in the field can be determined as follows:

- Field Water Efficiency (%) = (Water Transpired by Crop ÷ Water Applied to Field) x 100

Until 1960s, the common perception was that water was an infinite resource. At that time, there were fewer than half the current number of people on the planet. People were not as wealthy as today, consumed fewer calories and ate less meat, so less water was needed to produce their food. They required a third of the volume of water we presently take from rivers. Today, the competition for water resources is much more intense. This is because there are now more than seven billion people on the planet, their consumption of water-thirsty meat and vegetables is rising, and there is increasing competition for water from industry, urbanisation and biofuel crops. To avoid a global water crisis, farmers will have to strive to increase productivity to meet growing demands for food, while industry and cities find ways to use water more efficiently.

Successful agriculture is dependent upon farmers having sufficient access to water. However, water scarcity is already a critical constraint to farming in many parts of the world. With regards to agriculture, the World Bank targets food production and water management as an increasingly global issue that is fostering a growing debate. Physical water scarcity is where there is not enough water to meet all demands, including that needed for ecosystems to function effectively. Arid regions frequently suffer from physical water scarcity. It also occurs where water seems abundant but where resources are over-committed. This can happen where there is overdevelopment of hydraulic infrastructure, usually for irrigation. Symptoms of physical water scarcity include environmental degradation and declining groundwater. Economic scarcity, meanwhile, is caused by a lack of investment in water or insufficient human capacity to satisfy the demand for water. Symptoms of economic water scarcity include a lack of infrastructure, with people often having to fetch water from rivers for domestic and agricultural uses. Some 2.8 billion people currently live in water-scarce areas.

Technical Challenges

Irrigation schemes involve solving numerous engineering and economic problems while minimizing negative environmental impact.

- Competition for surface water rights.

- Overdrafting (depletion) of underground aquifers.

- Ground subsidence (e.g. New Orleans, Louisiana)

- Underirrigation or irrigation giving only just enough water for the plant (e.g. in drip line irrigation) gives poor soil salinity control which leads to increased soil salinity with consequent buildup of toxic salts on soil surface in areas with high evaporation. This requires either leaching to remove these salts and a method of drainage to carry the salts away. When using drip lines, the leaching is best done regularly at certain intervals (with only a slight excess of water), so that the salt is flushed back under the plant's roots.

- Overirrigation because of poor distribution uniformity or management wastes water, chemicals, and may lead to water pollution.

- Deep drainage (from over-irrigation) may result in rising water tables which in some instances will lead to problems of irrigation salinity requiring watertable control by some form of subsurface land drainage.

- Irrigation with saline or high-sodium water may damage soil structure owing to the formation of alkaline soil

- Clogging of filters: It is mostly algae that clog filters, drip installations and nozzles. UV and ultrasonic method can be used for algae control in irrigation systems.

Surface Irrigation

Furrow Irrigation of sugar cane in Australia, 2006

Surface irrigation is defined as the group of application techniques where water is applied and distributed over the soil surface by gravity. It is by far the most common form of irrigation throughout the world and has been practiced in many areas virtually unchanged for thousands of years.

Surface irrigation is often referred to as flood irrigation, implying that the water distribution is uncontrolled and therefore, inherently inefficient. In reality, some of the irrigation practices grouped under this name involve a significant degree of management (for example surge irrigation). Surface irrigation comes in three major types; level basin, furrow and border strip.

Drip irrigation involves enclosed passages, such as pipes and tubing.

Process

The process of surface irrigation can be described using four phases. As water is applied to the top end of the field it will flow or advance over the field length. The advance phase refers to that length of time as water is applied to the top end of the field and flows or advances over the field length. After the water reaches the end of the field it will either run-off or start to pond. The period of time between the end of the advance phase and the shut-off of the inflow is termed the wetting, ponding or storage phase. As the inflow ceases the water will continue to runoff and infiltrate until the entire field is drained. The depletion phase is that short period of time after cut-off when the length of the field is still submerged. The recession phase describes the time period while the water front is retreating towards the downstream end of the field. The depth of water applied to any point in the field is a function of the opportunity time, the length of time for which water is present on the soil surface.

Types of Surface Irrigation

Basin Irrigation

Level basin flood irrigation on wheat

Level basin irrigation has historically been used in small areas having level surfaces that are surrounded by earth banks. The water is applied rapidly to the entire basin and is allowed to infiltrate. In traditional basins no water is permitted to drain from the field once it is irrigated. Basin irrigation is favoured in soils with relatively low infiltration rates(Walker and Skogerboe 1987).This is also a method of surface irrigation. Fields are typically set up to follow the natural contours of the land but the introduction of laser levelling and land grading has permitted the construction of large rectangular basins that are more appropriate for mechanised broadacre cropping.

Drainback Level Basins

Drain back level basins (DBLB) or contour basins are a variant of basin irrigation where the field is divided into a number of terraced rectangular bays which are graded level or have no significant slope. Water is applied to the first bay (usually the highest in elevation) and when the desired depth is applied water is permitted to drain back off that bay and flow to the next bay which is at a lower elevation than the first. Each bay is irrigated in turn using a combination of drainage water from the previous bay and continuing inflow from the supply channel. Successful operation of these systems is reliant on a sufficient elevation drop between successive bays. These systems are commonly used in Australia where rice and wheat are grown in rotation.

Furrow Irrigation

Furrow irrigation is conducted by creating small parallel channels along the field length in the direction of predominant slope. Water is applied to the top end of each furrow and flows down the field under the influence of gravity. Water may be supplied using gated pipe, siphon and head ditch or bankless systems. The speed of water movement is determined by many factors such as slope, surface roughness and furrow shape but most importantly by the inflow rate and soil infiltration rate. The spacing between adjacent furrows is governed by the crop species, common spacings typically range from 0.75 to 2 metres. The crop is planted on the ridge between furrows which may contain a single row of plants or several rows in the case of a bed type system. Furrows may range anywhere from less than 100 m to 2000 m long depending on the soil type, location and crop type. Shorter furrows are commonly associated with higher uniformity of application but result in increasing potential for runoff losses. Furrow irrigation is particularly suited to broad-acre row crops such as cotton, maize and sugar cane. It is also practiced in various horticultural industries such as citrus, stone fruit and tomatoes.

Furrow irrigation system using siphon tubes

Gated pipe supply system

The water can take a considerable period of time to reach the other end, meaning water has been infiltrating for a longer period of time at the top end of the field. This results in poor uniformity with high application at the top end with lower application at the bottom end. In most cases the performance of furrow irrigation can be improved through increasing the speed at which water moves along the field (the advance rate). This can be achieved through increasing flow rates or through the practice of surge irrigation. Increasing the advance rate not only improves the uniformity but also reduces the total volume of water required to complete the irrigation.

Surge Irrigation

Surge Irrigation is a variant of furrow irrigation where the water supply is pulsed on and off in planned time periods (e.g. on for 1 hour off for 1½ hour). The wetting and drying cycles reduce infiltration rates resulting in faster advance rates and higher uniformities than continuous flow. The reduction in infiltration is a result of surface consolidation, filling of cracks and micro pores and the disintegration of soil particles during rapid wetting and consequent surface sealing during each drying phase. On those soils where surging is effective it has been reported to allow completion of the irrigation with a lower overall water usage and therefore higher efficiency and potentially offer the ability to practice deficit irrigation. The effectiveness of surge irrigation is soil type dependent; for example, many clay soils experience a rapid sealing behaviour under continuous flow and therefore surge irrigation offers little benefit.

Bay/Border Strip Irrigation

Border strip, otherwise known as border check or bay irrigation could be considered as a hybrid of level basin and furrow irrigation. The field is divided into a number of bays or strips, each bay is separated by raised earth check banks (borders). The bays are typically longer and narrower compared to basin irrigation and are orientated to align lengthwise with the slope of the field. Typical bay dimensions are between 10-70m wide and 100-700m long. The water is applied to the top end of the bay, which is usually constructed to facilitate free-flowing conditions at the downstream end. One common use of this technique includes the irrigation of pasture for dairy production.

Issues Associated with Surface Irrigation

While surface irrigation can be practiced effectively using the correct management under the right conditions, it is often associated with a number of issues undermining productivity and environmental sustainability:

- Waterlogging - Can cause the plant to shut down delaying further growth until sufficient water drains from the rootzone. Waterlogging may be counteracted by drainage, tile drainage or watertable control by another form of subsurface drainage.

- Deep drainage - Overirrigation may cause water to move below the root zone resulting in rising water tables. In regions with naturally occurring saline soil layers (for example salinity in south eastern Australia) or saline aqifers, these rising water tables may bring salt up into the root zone leading to problems of irrigation salinity.

- Salinization - Depending on water quality irrigation water may add significant volumes of salt to the soil profile. While this is a lesser issue for surface irrigation compared to other irrigation methods (due to the comparatively high leaching fraction), lack of subsurface drainage may restrict the leaching of salts from the soil. This can be remedied by drainage and soil salinity control through flushing.

The aim of modern surface irrigation management is to minimse the risk of these potential adverse impacts.

Drip Irrigation

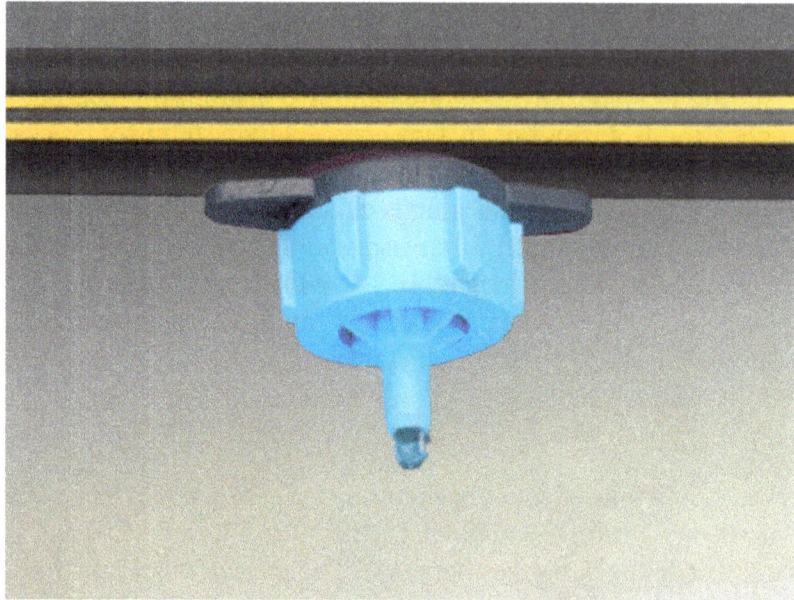

An Emitter or dripper in action

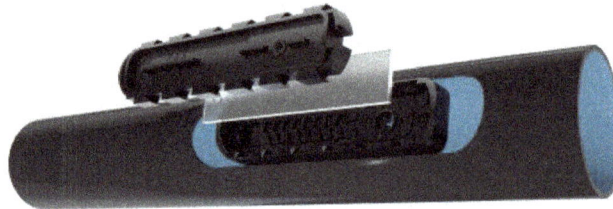

Open pressure compensated dripper

Drip irrigation is a form of irrigation that saves water and fertilizer by allowing water to drip slowly to the roots of many different plants, either onto the soil surface or directly onto the root zone, through a network of valves, pipes, tubing, and emitters. It is done through narrow tubes that deliver water directly to the base of the plant. It is chosen instead of surface irrigation for various reasons, often including concern about minimizing evaporation.

History

Primitive drip irrigation has been used since ancient times. Fan Sheng-Chih Shu, written in China during the first century BCE, describes the use of buried, unglazed clay pots filled with water as a means of irrigation. Modern drip irrigation began its development in Germany in 1860 when researchers began experimenting with subsurface irrigation using clay pipe to create combination irrigation and drainage systems. Research was later expanded in the 1920s to include the application of perforated pipe systems. The usage of plastic to hold and distribute water in drip irrigation was later developed in Australia by Hannis Thill.

Drip irrigation in Mexico vineyard, 2000

Usage of a plastic emitter in drip irrigation was developed in Israel by Simcha Blass and his son Yeshayahu. Instead of releasing water through tiny holes easily blocked by tiny particles, water was released through larger and longer passageways by using velocity to slow water inside a plastic emitter. The first experimental system of this type was established in 1959 by Blass who partnered later (1964) with Kibbutz Hatzerim to create an irrigation company called Netafim. Together they developed and patented the first practical surface drip irrigation emitter.

In the United States, the first drip tape, called *Dew Hose*, was developed by Richard Chapin of Chapin Watermatics in the early 1960s.

Modern drip irrigation has arguably become the world's most valued innovation in agriculture since the invention of the impact sprinkler in the 1930s, which offered the first practical alternative to surface irrigation. Drip irrigation may also use devices called micro-spray heads, which spray water in a small area, instead of dripping emitters. These are generally used on tree and vine crops with wider root zones. Subsurface drip irrigation (SDI) uses permanently or temporarily buried dripperline or drip tape located at or below the plant roots. It is becoming popular for row crop irrigation, especially in areas where water supplies are limited or recycled water is used for irrigation. Careful study of all the relevant factors like land topography, soil, water, crop and agro-climatic conditions are needed to determine the most suitable drip irrigation system and components to be used in a specific installation.

Components and Operation

Water distribution in subsurface drip irrigation

Nursery flowers watered with drip irrigation in Israel

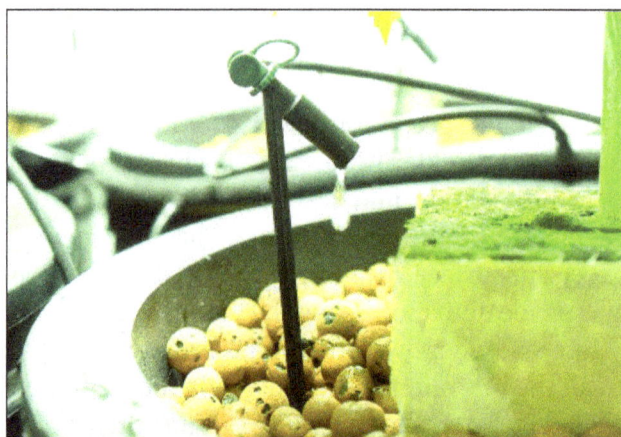

Horticulture drip emitter in a pot

Components used in drip irrigation (listed in order from water source) include:

- Pump or pressurized water source
- Water filter(s) or filtration systems: sand separator, Fertigation systems (Venturi injector) and chemigation equipment (optional)
- Backwash controller (Backflow prevention device)
- Pressure Control Valve (pressure regulator)
- Main line (larger diameter pipe and pipe fittings)
- Hand-operated, electronic, or hydraulic control valves and safety valves
- Smaller diameter polytube (often referred to as "laterals")
- Poly fittings and accessories (to make connections)
- Emitting devices at plants (emitter or dripper, micro spray head, inline dripper or inline driptube)

In drip irrigation systems, pump and valves may be manually or automatically operated by a controller.

Most large drip irrigation systems employ some type of filter to prevent clogging of the small emitter flow path by small waterborne particles. New technologies are now being offered that minimize clogging. Some residential systems are installed without additional filters since potable water is already filtered at the water treatment plant. Virtually all drip irrigation equipment manufacturers recommend that filters be employed and generally will not honor warranties unless this is done. Last line filters just before the final delivery pipe are strongly recommended in addition to any other filtration system due to fine particle settlement and accidental insertion of particles in the intermediate lines.

Drip and subsurface drip irrigation is used almost exclusively when using recycled municipal waste water. Regulations typically do not permit spraying water through the air that has not been fully treated to potable water standards.

Because of the way the water is applied in a drip system, traditional surface applications of timed-release fertilizer are sometimes ineffective, so drip systems often mix liquid fertilizer with the irrigation water. This is called fertigation; fertigation and chemigation (application of pesticides and other chemicals to periodically clean out the system, such as chlorine or sulfuric acid) use chemical injectors such as diaphragm pumps, piston pumps, or aspirators. The chemicals may be added constantly whenever the system is irrigating or at intervals. Fertilizer savings of up to 95% are being reported from recent university field tests using drip fertigation and slow water delivery as compared to timed-release and irrigation by micro spray heads.

Properly designed, installed, and managed, drip irrigation may help achieve water conservation by reducing evaporation and deep drainage when compared to other types of irrigation such as flood or overhead sprinklers since water can be more precisely applied to the plant roots. In addition, drip can eliminate many diseases that are spread through water contact with the foliage. Finally, in regions where water supplies are severely limited, there may be no actual water savings, but rather simply an increase in production while using the same amount of water as before. In very arid regions or on sandy soils, the preferred method is to apply the irrigation water as slowly as possible.

Pulsed irrigation is sometimes used to decrease the amount of water delivered to the plant at any one time, thus reducing runoff or deep percolation. Pulsed systems are typically expensive and require extensive maintenance. Therefore, the latest efforts by emitter manufacturers are focused toward developing new technologies that deliver irrigation water at ultra-low flow rates, i.e. less than 1.0 liter per hour. Slow and even delivery further improves water use efficiency without incurring the expense and complexity of pulsed delivery equipment.

An emitting pipe is a type of drip irrigation tubing with emitters pre-installed at the factory with specific distance and flow per hour as per crop distance.

An emitter restricts water flow passage through it, thus creating head loss required (to the extent of atmospheric pressure) in order to emit water in the form of droplets. This head loss is achieved by friction / turbulence within the emitter.

Advantages and Disadvantages

Drip irrigation and spare drip irrigation tubes in banana farm at Chinawal, India

Pot irrigation by On-line drippers

Pressure compensated integral dripper on soilless growing channels

The advantages of drip irrigation are:

- Fertilizer and nutrient loss is minimized due to localized application and reduced leaching.

- Water application efficiency is high if managed correctly

- Field levelling is not necessary.

- Fields with irregular shapes are easily accommodated.

- Recycled non-potable water can be safely used.

- Moisture within the root zone can be maintained at field capacity.

- Soil type plays less important role in frequency of irrigation.

- Soil erosion is lessened.

- Weed growth is lessened.

- Water distribution is highly uniform, controlled by output of each nozzle.

- Labour cost is less than other irrigation methods.

- Variation in supply can be regulated by regulating the valves and drippers.

- Fertigation can easily be included with minimal waste of fertilizers.

- Foliage remains dry, reducing the risk of disease.

- Usually operated at lower pressure than other types of pressurised irrigation, reducing energy costs.

The disadvantages of drip irrigation are:

- Initial cost can be more than overhead systems.

- The sun can affect the tubes used for drip irrigation, shortening their usable life. (This article does not include a discussion of the effects of degrading plastic on the soil content and subsequent effect on food crops. With many types of plastic, when the sun degrades the plastic, causing it to become brittle, the estrogenic chemicals (that is, chemicals replicating female hormones) which would cause the plastic to retain flexibility have been released into the surrounding environment.)

- If the water is not properly filtered and the equipment not properly maintained, it can result in clogging.

- For subsurface drip the irrigator cannot see the water that is applied. This may lead to the farmer either applying too much water (low efficiency) or an insufficient amount of water, this is particularly common for those with less experience with drip irrigation.

- Drip irrigation might be unsatisfactory if herbicides or top dressed fertilizers need sprinkler irrigation for activation.

- Drip tape causes extra cleanup costs after harvest. Users need to plan for drip tape winding, disposal, recycling or reuse.

- Waste of water, time and harvest, if not installed properly. These systems require careful study of all the relevant factors like land topography, soil, water, crop and agro-climatic conditions, and suitability of drip irrigation system and its components.

- In lighter soils subsurface drip may be unable to wet the soil surface for germination. Requires careful consideration of the installation depth.

- most drip systems are designed for high efficiency, meaning little or no leaching fraction. Without sufficient leaching, salts applied with the irrigation water may build up in the root zone, usually at the edge of the wetting pattern. On the other hand, drip irrigation avoids the high capillary potential of traditional surface-applied irrigation, which can draw salt deposits up from deposits below.

- the PVC pipes often suffer from rodent damage, requiring replacement of the entire tube and increasing expenses.

- Drip irrigation systems cannot be used for damage control by night frosts (like in the case of sprinkler irrigation systems)

Uses

Irrigation dripper

Drip irrigation is used in farms, commercial greenhouses, and residential gardeners. Drip irrigation is adopted extensively in areas of acute water scarcity and especially for crops and trees such as coconuts, containerized landscape trees, grapes, bananas, pandey, eggplant, citrus, strawberries, sugarcane, cotton, maize, and potatoes.

Drip irrigation for garden available in drip kits are increasingly popular for the homeowner and consist of a timer, hose and emitter. Hoses that are 4 mm in diameter are used to irrigate flower pots.

Rainfed agriculture

The term rainfed agriculture is used to describe farming practises that rely on rainfall for water. It provides much of the food consumed by poor communities in developing countries. For example, rainfed agriculture accounts for more than 95% of farmed land in sub-Saharan Africa, 90% in Latin America, 75% in the Near East and North Africa; 65% in East Asia and 60% in South Asia.

Levels of productivity, particularly in parts of sub-Saharan Africa and South Asia, are low due to degraded soils, high levels of evaporation, droughts, floods and a general lack of effective water management. A major study into water use by agriculture, known as the Comprehensive Assessment of Water Management in Agriculture, coordinated by the International Water Management Institute, noted a close correlation between hunger, poverty and water. However, it concluded that there was much opportunity to raise productivity from rainfed farming.

The authors considered that managing rainwater and soil moisture more effectively, and using supplemental and small-scale irrigation, held the key to helping the greatest number of poor people. It called for a new era of water investments and policies for upgrading rainfed agriculture that would go beyond controlling field-level soil and water to bring new freshwater sources through better local management of rainfall and runoff.

The importance of rainfed agriculture varies regionally but produces most food for poor communities in developing countries. In subSaharan Africa more than 95% of the farmed land is rainfed, while the corresponding figure for Latin America is almost 90%, for South Asia about 60%, for East Asia 65% and for the Near East and North Africa 75%. Most countries in the world depend primarily on rainfed agriculture for their grain food. Despite large strides made in improving productivity and environmental conditions in many developing countries, a great number of poor families in Africa and Asia still face poverty, hunger, food insecurity and malnutrition where rainfed agriculture is the main agricultural activity. These problems are exacerbated by adverse biophysical growing conditions and the poor socioeconomic infrastructure in many areas in the semi-arid tropics (SAT). The SAT is the home to 38% of the developing countries' poor, 75% of whom live in rural areas. Over 45% of the world's hungry and more than 70% of its malnourished children live in the SAT

There is a correlation between poverty, hunger and water stress. The UN Millennium Development Project has identified the 'hot spot' countries in the world suffering from the largest prevalence of malnourishment. These countries coincide closely with those located in the semi-arid and dry subhumid hydroclimates in the world (Fig. 1.1), i.e. savannahs and steppe ecosystems, where rainfed agriculture is the dominating source of food and where water constitutes a key limiting factor to crop growth. Of the 850 million undernourished people in the world, essentially all live in poor, developing countries, which predominantly are located in tropical regions.

Since the late 1960s, agricultural land use has expanded by 20–25%, which has contributed to approximately 30% of the overall grain production growth during the period. The remaining yield outputs originated from intensification through yield increases per unit land area. However, the regional variation is large, as is the difference between irrigated and rainfed agriculture. In developing countries rainfed grain yields are on average 1.5 t/ha, compared with 3.1 t/ha for irrigated yields (Rosegrant et al., 2002), and increase in production from rainfed agriculture has mainly originated from land expansion. Trends are clearly different for different regions. With 99% rainfed production of main cereals such as maize, millet and sorghum, the cultivated cereal area in sub-Saharan Africa has doubled since 1960 while the yield per unit of land has been nearly stagnant for these staple crops (FAOSTAT, 2005). In South Asia, there has been a major shift away from more drought-tolerant, low-yielding crops such as sorghum and millet, while wheat and maize hasapproximately doubled in area since 1961 (FAOSTAT, 2005). During the same period, the yield per unit of land for maize and wheat has more than doubled (Fig. 1.2). For predominantly rainfed systems, maize crops per unit of land have nearly tripled and wheat more than doubled during the same time period. Rainfed maize yield differs substantially between regions (Fig. 1.2a). In Latin America (including the Caribbean) it exceeds 3 t/ha, while in South Asia it is around 2 t/ha and in subSaharan Africa it only just exceeds 1 t/ha. This can be compared with maize yields in the USA or southern Europe, which normally amount to approximately 7–10 t/ha (most maize in these regions is irrigated). The average regional yield per unit of land for wheat in Latin America (including the Caribbean) and South Asia is similar to the average yield output of 2.5–2.7 t/ha in North America (Fig. 1.2b). In comparison, wheat yield in Western Europe is approximately twice as large (5 t/ha), while in sub-Saharan Africa it remains below 2 t/ha. In view of the historic regional difference in development of yields, there appears to exist a significant potential for raised yields in rainfed agriculture, particularly in sub-Saharan Africa and South Asia.

Rural development through sustainable management of land and water resources gives a plausible solution for alleviating rural poverty and improving the livelihoods of the rural poor. In an effective convergence mode for improving the rural livelihoods in the target districts, with watersheds as the operational units, a holistic integrated systems approach by drawing attention to the past experiences, existing opportunities and skills, and supported partnerships can enable change and improve the livelihoods of the rural poor. The well-being of the rural poor depends on fostering their fair and equitable access to productive resources. The rationale behind convergence through watersheds has been that these watersheds help in 'cross-learning' and drawing on a wide range of experiences from different sectors. A significant conclusion is that there should be a balance between attending to needs and priorities of rural livelihoods and enhancing positive directions of change by building effective and sustainable partnerships. Based on the experience and performance of the existing integrated community watersheds in different socioeconomic environments, appropriate exit strategies, which include proper sequencing of interventions, building up of financial, technical and organizational capacity of local communities to internalize and sustain interventions, and the requirement for any minimal external technical and organizational support needs to be identified.

While absolute grain yield variations exist between different global locations as cited in this article, the potential for improved rain fed grain yields may be less than is suggested by a comparison between sub-Saharan and European locations for example, this applies particularly in areas where grain yield is primarily determined by the growing season rainfall. A more accurate formula for measuring yield potential is y x a = X, where X is grain yield in Kg/hectare, y is millimetres of growing season rainfall and a is the variable yield factor, a number that may vary somewhere between 5 and 20. Thus assuming a variable yield factor of 15 and a growing season rainfall of 220mm the formula would express as 220mm x 15 = 3300 kg/ha yield potential. This formula, while not taking account of either the carryover benefits of stored rainfall in the soil profile resulting from out of season rainfall or the impact of temperature or soil fertility, still gives a more accurate picture of the degree to which actual grain yields are matching the region's potential yields and is a better basis for comparison between very different regions such as Europe and the sub-Sahara. A European yield of 5000 kg/ha from a rainfall of 500mm results in a variable yield factor of 10 while a yield of 2500 kg/ha in a 200mm rainfall area has a variable yield factor of 12.5, in such an example the lower yielding crop has actually been 25% more efficient in its rainfall utilisation than the higher yielding crop. Agronomy needs to be based on attaining the highest possible variable yield factor rather than the highest absolute yield, factor numbers as high as 17+ are achievable.

References

- G. Mokhtar (1981-01-01). Ancient civilizations of Africa. Unesco. International Scientific Committee for the Drafting of a General History of Africa. p. 309. ISBN 9780435948054. Retrieved 2012-06-19.

- Richard Bulliet, Pamela Kyle Crossley, Daniel Headrick, Steven Hirsch. Pages 53-56 (2008-06-18). The Earth and Its Peoples, Volume I: A Global History, to 1550. Books.google.com. ISBN 0618992383. Retrieved 2012-06-19.

- Frenken, K. (2005). Irrigation in Africa in figures – AQUASTAT Survey – 2005 (PDF). Food and Agriculture Organization of the United Nations. ISBN 92-5-105414-2. Retrieved 2007-03-14.

- Drainage Manual: A Guide to Integrating Plant, Soil, and Water Relationships for Drainage of Irrigated Lands. Interior Dept., Bureau of Reclamation. 1993. ISBN 0-16-061623-9.

- R. Goyal, Megh (2012). Management of drip/trickle or micro irrigation. Oakville, CA: Apple Academic Press. p. 104. ISBN 978-1926895123.

- Drainage Manual: A Guide to Integrating Plant, Soil, and Water Relationships for Drainage of Irrigated Lands. Interior Dept., Bureau of Reclamation. 1993. ISBN 0-16-061623-9.

- "Africa, Emerging Civilizations In Sub-Sahara Africa. Various Authors; Edited By: R. A. Guisepi". History-world.org. Retrieved 2012-06-19.

Permissions

Index